AF880738

Laboratories in Mathematical Experimentation: A Bridge to Higher Mathematics 2nd Edition

J. William Bruce[1]
George Cobb[2]
Giuliana Davidoff[2]
Christopher Dugaw[3]
Alan Durfee[2]
Art M. Duval[4]
Janice Gifford[2]
Helmut Knaust[4]
Donal O'Shea[2]
Mark Peterson[2]
Harriet Pollatsek[2]
Margaret Robinson[2]
Lester Senechal[2]
Robert Weaver[2]

1. University of Liverpool, Liverpool, United Kingdom
2. Mount Holyoke College, South Hadley, Massachusetts
3. California Polytechnic University, Humboldt, Arcata, California
4. The University of Texas at El Paso, El Paso, Texas

THE PRESS AT CAL POLY HUMBOLDT

This book is dedicated to Elizabeth Bergmann Hutchcroft, Mount Holyoke College class of 1935, and to the memory of her late husband, C. Robert Hutchcroft. Mrs. Hutchcroft, a mathematics major at Mount Holyoke, earned a master's degree in mathematics at Teachers' College, Columbia University, and went on to teach mathematics at her high school alma mater, *the Lincoln School. She has established the Robert and Elizabeth Bergmann Hutchcroft Endowment Fund to advance the study of mathematics at Mount Holyoke. The Fund supports curricular innovation, faculty research leaves, faculty development, and necessary equipment and technology for the department. Mrs. Hutchcroft's generosity ensures that projects like the Laboratory in Mathematical Experimentation and this text will continue to flourish at Mount Holyoke and enrich the mathematics education of our students. We are profoundly grateful.*

The mathematicians and statisticians at Mount Holyoke College

South Hadley, 1996

©2023 J. William Bruce, George Cobb, Giuliana Davidoff, Christopher Dugaw, Alan Durfee, Art M. Duval, Janice Gifford, Helmut Knaust, Donal O'Shea, Mark Peterson, Harriet Pollatsek, Margaret Robinson, Lester Senechal, and Robert Weaver

This book has been made available under a Creative Commons Attribution-Non-Commercial 4.0 International (CC BY-NC 4.0).
For details, please refer to creativecommons.org/licenses/by-nc/4.0/legalcode.

The Press at Cal Poly Humboldt showcases Cal Poly Humboldt research and scholarship with a broad scope of print and electronic publications including books, journals, conference proceedings, data sets, open textbooks, and more. The open-access works connect Cal Poly Humboldt with a worldwide community of researchers and scholars.

The Press at Cal Poly Humboldt
California State Polytechnic University, Humboldt
University Library
1 Harpst Street
Arcata, California 95521-8299
press@humboldt.edu
press.humboldt.edu

ISBN: 978-1-962081-11-5

Cover shows visualization of 3-adic numbers, design by Sarah Godlin

Preface

The mathematics department at Cal Poly Humboldt recently underwent a complete major redesign with a desire to create a more active and engaging curriculum. The goal was to embrace the motto of the California Polytechnic system "learn by doing." The search for a model sophomore-level bridge course led us to this text. The faculty at Mt. Holyoke College were extremely generous in sharing their knowledge and experience teaching this course, especially Margaret Robinson. The only notable challenge to using the text was the lack of modern computer code. Thus, I sought to create this new edition which incorporates code written for *Matlab* or its open-source clone *Octave*, as well as code written in *Mathematica* by Helmut Knaust and Art Duval at The University of Texas at El Paso where they also teach a course using this text.

The changes in this new edition were intentionally kept to a minimum to maintain the character and spirit of the original text. I have updated the pseudocode to be more consistent with modern coding and added the new code to the end of the chapters where appropriate. All of the *Mathematica* code and corresponding explanations were written by Art Duval and Helmut Knaust. The 1st edition used a program called *Calcwin* to create certain plots and perform other calculations, but it does not easily run on modern computers. This edition includes instructions for using Desmos and/or *Mathematica* code to accomplish everything that was previously done using *Calcwin*.

I am truly grateful to the original authors of this text and all that supported its creation. I would also like to thank my colleagues in the Mathematics Department at Cal Poly Humboldt for their contributions to our curriculum redesign and especially Peter Goetz, Kamila Larripa, and Walden Freedman who worked on this course. I thank the Center for Teaching and Learning at Cal Poly Humboldt for supporting our curriculum development, and the university for providing me a sabbatical which allowed me to complete this new edition.

Christopher Dugaw, Cal Poly Humboldt

Arcata, CA, 2024

Preface to the 1st edition

In 1989, after much discussion, we added a new sophomore-level "bridge" course to our curriculum. Just as it is difficult to learn to write unless you have something you want to say, so too, we reasoned, it is difficult to learn to construct a proof unless you have something you want to prove. We wanted to share with our students the joy, and the frustration, of mathematical discovery. We wanted to immerse them in exploration of mathematical phenomena, to have them discover things for themselves, to have them make conjectures and construct arguments in support of those conjectures. We wanted them to discover things that they understood might well be false and which, therefore, needed proof. We wanted them to encounter a broad range of mathematical phenomena early in their mathematical careers and to take with them a solid base of experience on which to build future understanding.

Our course, which we called the *Laboratory in Mathematical Experimentation* and which students called "the Lab", succeeded beyond any of our expectations. After just one year, we found students much more likely to read mathematics actively, more likely to dive in and "mess around" with a hard problem, more likely to ask questions and look for patterns, and more likely to formulate an argument clearly. Students who have taken the Lab do better in the real analysis and abstract algebra courses required for the mathematics major than those who have not. Moreover, both students and faculty enjoy the course. We find it easy to teach: reading papers and grading *is* onerous, but preparation is trivial and office hour time is modest. Students enjoy setting their own agendas and inevitably become caught up in their investigations. The atmosphere of collegiality and shared inquiry sharpens students' interest in mathematics and helps them think of themselves as mathematicians. We now require the course of every mathematics major.

This book grows out of the Lab course and consists of a collection of sixteen laboratory investigations in mathematics accessible to beginning college students. Each investigation invites students to observe and to look for patterns and encourages them to establish language to describe, conjecture, and analyze the phenomena under study. Each investigation leads to mathematics which students will encounter in later courses and seeks to supply the student with a repertoire of concrete examples to nourish their intuition. Most will also result in the student discovering some things she believes to be true and wants to prove, but cannot. In a typical offering of the Lab, students do six or seven of the investigations.

This book could be used to offer a course like ours. It could also be used to supplement other courses or as a source for students' independent projects. The sixteen investigations are almost all independent of each other, and most do not require calculus (the exceptions to both statements are noted in the Introduction). All but two of the investigations require a computer (or programmable calculator). The accompanying instructor's manual says more about each of the investigations. It also gives much more detail about the Lab course.

This work is truly a collective effort. Every mathematician and statistician in our department has had a hand in shaping it and the course from which it grew. When it comes to the Lab, we also consider J. William Bruce of the University of Liverpool to be an honorary department member, since while on sabbatical at Mount Holyoke he taught the Lab, made important contributions to every chapter he used, and contributed several additional projects.

It is a pleasure to acknowledge the support of the National Science Foundation, first for laboratory computers, then for curriculum development and writing, and finally for dissemination through their Undergraduate Faculty Enhancement program. We are also grateful for NSF's insistence that we form an Advisory Board for this project. Advice, suggestions and encouragement from Thomas Cecil (College of the Holy Cross), Gregory Fredricks (Lewis and Clark College), Gregory Hill (University of North Texas), and Jacob Sturm (Rutgers University, Newark) have been invaluable, and we thank them. We also gratefully acknowledge support from the Sloan Foundation and Hewlett Packard.

We also thank several other colleagues. Mizan Kahn, now at Eastern Connecticut University, taught and commented on an early version of the Lab. More recently Patrick Fitzpatrick, a sabbatical visitor from University College Cork, taught the Lab and contributed many ideas, including the "warm-up" exercise on diagonals of rectangles. Our students have also helped us in many ways, especially the junior and senior majors in the fall of 1988 who helped design and test the projects used in the very first offering of the Lab: Tessa Campbell, Julie Derynda, Barbara Hswe, Kristine Kusek, Kathleen Malone, and Ke Wu.

We received many valuable suggestions from the participants in our NSF-UFE workshop in June 1996, as we were preparing the final version of this text. We thank Mysore Jagadish and Pedro Suarez (Barry University), Olusola Akinyele (Bowie State University), Terrence Bisson and Donald Girod (Canisius College), Barbara Reynolds (Cardinal Stritch College), Alan Levine and Ben Shanfelder (student) (Franklin and Marshall College), John Kellett (Gettysburg College), Lynnell Matthews (Howard Community College), Kathy Kraft and Robert Woodle (Jamestown College), Douglas Burkholder and Mary Flagg (McPherson College), Arup Mukherjee and Ethel Wheland (Pennsylvania State University), Steve Cohen and John Currano (Roosevelt University), Donald Miller (Saint Mary's College), Patricia Army and Barbara Becker (Saint Xavier University), Craig Bailey and Charles Hanna (U.S. Naval Academy), Bruce Lundberg (University of Southern Colorado), Jiu Ding (University of Southern Mississippi), Michael Evans (Washington and Lee University), Karl David and Arnold Shilepsky (Wells College), and Mark Janeba and Frank Zizza (Willamette University).

Finally we thank our colleagues in other departments at Mount Holyoke and our spouses for their support and encouragement.

The mathematicians and statisticians at Mount Holyoke College

South Hadley, 1996

Contents

Introduction

You undoubtedly have been told that the best way to learn mathematics is to do mathematics. But what does it mean to *do mathematics*? To be most people, doing mathematics is working out isolated problems in textbooks. But to those who love mathematics, doing mathematics is the process of exploring various mathematical phenomena. It is about investigation, discovery, mystery and understanding. It is a process where finding an illuminating way to think about something leads to more questions and more possibilities, where elementary notions turn out to be unexpectedly subtle, and where seemingly intractable difficulties sometimes yield totally to a new idea. This book consists of a series of laboratory projects, most of which use the computer as an experimental tool. For each project, your tasks will be

- To work by hand and/or by computer to generate examples illuminating questions asked—and to raise some questions of your own;
- To carry out suitable experiments to enable you to see *patterns* in the data relating to the problem under investigation;
- To give clear descriptions of your experimental findings;
- To make *conjectures* based on your observations;
- To support your conjectures with arguments based on your emprical evidence, on mathematical analysis, and—when possible—with *mathematical proofs*.

The topics range across many different areas of mathematics and statistics. We have chosen them both to convey some of the breadth of the mathematical sciences and also to introduce you to a number of important ideas that you will encounter again in future courses. With a few exceptions described below, the chapters (projects) can be covered in any order. You can have a valuable learning experience working through one or two or half a dozen or more.

The goal is for you to have the fun of discovering some mathematics on your own. While you will learn some specific ideas and techniques, these are secondary to the broader experience of mathematical inquiry. Of course, what distinguishes mathematics from other scientific disciplines is the possibility that empirical findings can be rigorously *proved* beyond a shadow of a doubt. You won't always know enough mathematics to construct such a proof, but you can create a good foundation for learning to construct proofs in more advanced courses by formulating your empirical arguments rigorously and clearly.

Each chapter introduces a topic and places it in some context in the field. There may be some *exercises* to give you practice with new ideas. It then raises a number of more substantial *questions* for you to investigate, with some suggestions for how to get started. The chapter concludes with a discussion of some of the underlying mathematical ideas, to help you understand and interpret your results and to give you ideas for how to support some of your conjectures with analysis or, in some cases, proof.

You will find it useful to keep a laboratory notebook of all your experiments, jotting things down as you do them and recording your observations and guesses. Your notebooks can be the basis for discussion with fellow students as well as for writing a report summarizing the results of your investigation.

Writing a report is an invaluable opportunity to clarify and refine your thinking. Your instructor may specify which questions your report should address, or you may choose a cluster of related questions which interest you. We suggest that you write your report so that it makes sense to a reader who has taken a semester of college level mathematics but has not worked with this material. (If you have a friend who fits this description and is willing to read and comment on your drafts, you have a treasure!) You should write in full sentences and paragraphs—no cryptic strings of formulas. Try to be both clear and interesting. Look at a math text you've particularly liked or an article you enjoyed reading to get an idea of the tone and style to aim for.

Your introduction should describe the topic under investigation in a way that engages the reader's interest. You may need to provide some background or context for your investigation. Define with care the terminology that you will use, since precise descriptions of the phenomena you observe are essential. (Sometimes it is easiest to write the introduction last!) The body of your report naturally falls into four sections.

1. Your experimental strategy or *design*.

   You should motivate the questions you ask—and the order in which you ask them—and explain the logic of your choice of examples. Here are some specific suggestions to get you started.

   (a) Describe your first example.
      - What was it?
      - Describe it geometrically and/or algebraically.
      - Why did you choose it?
      - What happened when you carried it out?

   (e) What did you try next? Why? What were your results?

   (f) What eventually evolved as your general strategy for choosing examples? Why?

2. Results of your experimentation.

   You should organize your data carefully and give thought to how you display your results; make effective use of tables, graphs, and pictures.

   (a) Describe how your various examples worked out, being as clear as you can, but omitting details that don't seem important.

   (b) Attach tables or graphs or sketches to your description, where appropriate. Give each a clear, informative title.
   However, don't include anything you don't refer
   to in your discussion section.

3. Analysis of data.

   Organize the discussion of your data carefully, and refer to your results by citing the titles and numbers you assign. (e.g., a report on Chapter 4 might refer to: "Table 3, Mersenne primes.") Explain how your data support your conjectures.

(a) What patterns do you observe in your data?

(b) Formulate your conjectures. Which patterns do you guess represent *real* phenomena, rather than accidental regularities of the examples you happened to choose?

(c) Justify your conjectures. Your choice of examples should test your conjectures stringently—try to rule out "chance" regularities.

4. Mathematical analysis of conjectures.

   Back up your empirical argument with an analytical and/or theoretical one when you can.

You will need access to a computer, of course. Most chapters require one or more simple computer programs. Programs are described in the text in *pseudocode*, an outline of the program that makes its logic clear without burdening the reader with the details of the syntax of any particular programming language. Working code for each program is usually provided at the end of the chapter in two languages, *MATLAB/Octave* and *Mathematica.* The *MATLAB/Octave* code is written so that it will run in *MATLAB*, or its free open source alternative *Octave* which can be downloaded from: https://octave.org/. Three of the chapters (5, 11 and 12) use software that is free for use on the web, (e.g., *Desmos*). We do not assume prior knowledge of programming, and learning to program is not one of the goals of this text. However, by trying to *read* the programs and understand what they do, you will learn something about what computers can do and how to make them do it.

The latest code is available for download from Github. The *MATLAB/Octave* code can be found at:

https://github.com/cdugaw/Laboratories-in-Mathematical-Experimentation-Matlab-Octave

and *Mathematica* code can be found here:

https://github.com/HKnaust/Laboratories-in-Mathematical-Experimentation

To download the code click the green button labeled **<>Code** and select **Download Zip**.

Most of the chapters are completely independent of each other and of specific prior courses, but there are some exceptions. Chapters 13, 14 and 15, Iteration to Solve Equations, Iteration of Quadratic Functions, and Iteration of Linear Maps in the Plane are independent of each other, but they each presuppose that you have worked through Chapter 1, Iteration of Linear Functions. Chapter 16, the Euclidean Algorithm for the Complex Integers, assumes you have done Chapter 3 on the Euclidean Algorithm.

Most of this text doesn't require calculus. The derivative appears briefly, but avoidably, in portions of Chapters 12 and 14. There is modest use of the definite integral in Chapter 10, Numerical Integration, and a portion of Chapter 11, Sequences and Series. However, the investigations in Chapter 10 are mostly self-contained and actually provide a good introduction to the integral. Integration and the fundamental theorem are used more heavily in Chapter 12, Experiments in Periodicity.

We hope you enjoy exploring these projects and that they whet your appetite for further study.

Chapter 1

Iteration of Linear Functions

1.1 Introduction

This chapter introduces iteration of linear functions, an interesting topic in its own right, and a prelude to other labs in which iteration is explored in greater depth. The computer provides an especially appropriate tool for the study of iteration since it is so adroit at doing repetitive operations. Sequences of numbers arise from the iteration process. In some cases, the sequences will converge, while in others they will diverge in a variety of ways.

In this chapter you will:

- Learn what iteration and iteration sequences are;
- See examples of iteration sequences that converge and diverge;
- Look for examples of linear functions producing convergent and divergent iteration sequences; and
- Determine general conditions on linear functions predicting whether their iteration sequences converge or diverge.

1.2 What is iteration?

Have you ever idly repeatedly pressed one of the buttons on your calculator? Consider for example the $\sqrt{\ }$ button. You could compute

$$\sqrt{2}, \sqrt{\sqrt{2}}, \sqrt{\sqrt{\sqrt{2}}}, \sqrt{\sqrt{\sqrt{\sqrt{2}}}}, \ldots$$

successively by entering the number 2 and pressing the $\sqrt{\ }$ button repeatedly. This is an example of *iteration*, that is, repeated application of the same function given some initial value. Numbers obtained in this way are called *iterates* of the starting value.

If you carry out the process above you will be rewarded with a sequence of numbers

$$1.414\ldots, 1.189\ldots, 1.090\ldots, 1.044\ldots, 1.021\ldots, 1.010\ldots, \ldots$$

which are getting nearer and nearer to 1. We say that the sequence is *converging* to 1. Indeed, you will find the same behavior if you start with any other positive number. For example, beginning instead with 5, you obtain the sequence

$$2.236\ldots, 1.495\ldots, 1.222\ldots, 1.105\ldots, 1.051\ldots, 1.025\ldots, \ldots$$

approaching 1 from above, and beginning with 0.2, you will get

$$0.447\ldots, 0.669\ldots, 0.818\ldots, 0.904\ldots, 0.951\ldots, 0.975\ldots, \ldots$$

which approaches 1 from below.

In Chapter 14, we find that even a simple function $f(x)$ can give rise to a highly complicated sequence of iterates

$$x_0,\ x_1 = f(x_0),\ x_2 = f(x_1),\ x_3 = f(x_2),\ \ldots$$

Indeed iteration is very much a field of active current research. Iteration is also of some practical importance. Your education to date has been rather misleading. You may think that most equations can be solved *exactly*. Nothing could be further from the truth! Most equations can only be solved approximately, and, as we may show in Chapter 13, iteration is a useful method for obtaining ever more accurate approximate solutions. Chapter 15 uses iteration as a problem-solving tool in yet another setting.

1.3 The mathematical ideas

In this lab, we will ask the computer to do most of the boring work (the computations) and leave the interesting analysis for us to do. We shall use a program, ITERLIN, that computes n iterates of a function $f(x) = ax + b$ for given values of a and b, and for a given starting point $x = x_0$. The algorithm can be described by the following program outline or, as it's sometimes called, *pseudocode*. Using pseudocode allows us to show the logic of an algorithm without requiring knowledge of the details of any specific computer language. (Actually, our program outlines are somewhat more detailed than standard pseudocode, since they're designed for novice readers of programs. We want you—armed with a manual for your programming language—to be able to create a working program from the pseudocode.) In this first example of pseudocode, the lines are numbered so that we can comment on them one by one.

Program outline: Iterlin

```
1 Input:   coefficients a and b of a linear function
           y = ax + b, an initial value x0, and the
           number of iterations n
2 Output:  the n iterates of the function, starting with x0

3   INITIALIZE x as an n-dimensional vector
4   x(1) := x0           %set first entry of x to x0
5   FOR i = 1 TO n-1
6        x(i+1) := a x(i) + b   %iterate
7   END FOR
8   OUTPUT X
```

1. Most programs require the user to provide some information (input). Because the input is specified here, we don't repeat input statements in the program outline that follows (lines 3 to 8).

2. The only way the user knows the result of the computer's calculations is if the values of interest are printed or graphically displayed. Because the form of the output can vary, we *do* repeat an output statement in the body of the pseudocode (line 8).

3. In this line we need to tell the computer that x as an n-dimensional vector which will later be filled with the iterates (see lines 4 and 7).

4. We assign to the first entry of the vector x to the initial value $x0$. Notice that in this pseudocode the first entry of the vector x is x(1). In MATLAB/Octave the first entry of a vector is addressed with a 1, while in some other languages the first entry of a vector is addressed with a 0. The symbol ":=" denotes *assignment*. The expression "C := D", means that the variable C is given the value D:

$$(\text{new value of } C) \;=\; (\text{current value of } D).$$

We write "x(1) := x0" to indicate that the value assigned to the first entry of x is the input value x_0. Note that there is no way to type subscripts in computer languages, so we type "x0."

For another example, to indicate that the value of the variable i is to be increased by one from its current value, we write "i := i + 1, "

$$(\text{new value of i}) \;=\; (\text{old value of i}) + 1.$$

With the usual meaning of equality, the equation $i = i + 1$ doesn't make sense, since there is no value of i that makes the equation true. Computer languages (i.e., real code, not pseudocode) differ in how they represent assigment. *Mathematica* uses "C := D" while *Matlab/Octave* use "C = D, " to name two examples. So don't necessarily expect to see ":=" in a working program. We have used the distinctive ":=" to draw your attention to the fact that this isn't the equality you're used to. (Some authors use "C ← D" to denote assignment in pseudocode.)

Regular equality *is* used in computer programs as well. Equality is used in *logical* statements: statements which the computer must test for truth or falsehood. For example, if you only wanted to see values of x which make $y = ax + b$ equal to zero, you would write "IF y = 0 THEN PRINT x."

Sometimes our pseudocode includes remarks preceded by the symbol "%". These *comments*, as they are called, are to be read by the human user; they are invisible to the computer. Most computer languages have a special symbol like "%" that can be used to add comments for the reader in this way. Comments are essential in complex programs. In this case the comment tells us that this line assigns the value $x0$ to the first entry of the vector x.

5. The heart of this algorithm is what's called a *FOR loop*. It is traditional to indent the body of the loop (line 6) to make it easy to identify. This loop begins with $i = 1$. You can think of i as a counter keeping track of how many times the computer has gone through the loop. With $i = 1$, it is making its first pass through the loop.

6. We write "x(i+1) := a x(i) + b" to indicate that the value assigned to the $i+1$ entry of the vector x is the result of the calculation "a x(i) + b". Note that the first time this arithmetic is done (the passage through the loop with $i = 1$), it is done with with the values of a, b, and the 1st entry of x that was set in line 4.

7. The "END FOR" statement, in conjunction with line 5, means: check to see if $i > n - 1$, and if so, stop; if not, increase i by 1 (in computerese, set i := i + 1) and repeat the loop. Notice that on the *second* pass through the loop, $i = 2$ and the statement in line 6 will set the 3rd entry of the vector x, $x_2 = f(x_1)$. At the end of the second time through the loop, i will become 3, and so forth.

Read through the pseudocode and be sure you understand every line. (A good check of your understanding is to choose some values of a, b, x_0 and a small n and follow the computer instructions yourself.) Working programs in *Matlab/Octave* and *Mathematica* derived from this outline appear at the end of this chapter. They perform the desired iterations. You may use the programs we give or write your own in whatever language you wish.

Practice running your version of ITERLIN with various values of a, b, x_0, and n. For instance, let $f(x) = -2x + 1$ with the initial $x = 1.5$ using 10 iterations. We will use the notation $(a, b, x_0, n) = (-2, 1, 1.5, 10)$ for this experiment. Also try $(a, b, x_0, n) = (0.5, 2, 5, 10)$. These two examples lead to rather different outcomes. Describe them as carefully as you can. Try also the examples $(a, b, x_0, n) = (-3, 1, 1, 15)$ and $(a, b, x_0, n) = (-3, 1, 0.25, 15)$. What do you notice about these last two experiments, which both use the same linear function? Describe these outcomes carefully too. How are they similar to the results of the first two? How are they different?

The graphs of functions of the form $f(x) = ax + b$, where a and b are particular constants, are straight lines in the plane (and hence f is called a *linear* function). In this project we will explore the behavior of linear functions under iteration. So, given a linear function $f(x) = ax + b$ and some initial choice for x, say x_0, we produce a sequence of numbers $x_1, x_2, \ldots$ defined by

$$\begin{aligned} x_1 &= ax_0 + b \\ x_2 &= ax_1 + b \\ x_3 &= ax_2 + b \\ &\vdots \\ x_{n+1} &= ax_n + b \\ &\vdots \end{aligned}$$

We shall be especially interested in how this sequence behaves as n gets larger and larger. Clearly the behavior of the sequence $\{x_n\}$ depends on the linear function under consideration (that is, the choice of a and b) as well as the choice of initial starting point x_0. We wish to investigate how the various choices influence the eventual behavior of the sequence. The computer is an ideal tool to use to explore such questions.

In order to keep track of patterns, we need some language to describe the behavior we see. Mathematicians distinguish between *convergent* sequences—like those suggested by the experiments $(a, b, x_0, n) = (0.5, 2, 5, 10)$ and $(a, b, x_0, n) = (-3, 1, 0.25, 15)$—and *divergent* sequences—like those suggested by the experiments $(-2, 1, 1.5, 10)$ and $(-3, 1, 1, 15)$. Convergent sequences *converge to* a particular value, called the *limit* of the sequence. The experiment $(a, b, x_0, n) = (0.5, 2, 5, 10)$ suggests the limit is 4; and the experiment $(a, b, x_0, n) = (-3, 1, 0.25, 15)$ suggests the limit is 0.25.

Exercise: Write down careful definitions of the three terms *convergent sequence*, *divergent sequence*, and *limit* of a convergent sequence.

1.4 Questions to explore

For each of the questions below, use the computer program to find your own examples of linear functions, different from those you do in class, which have the specified behavior and help to shed light on the posed question:

Question 1: Can you find a linear function that gives a convergent iteration sequence of values for *every* initial value? That is, can you find values of a and b that assure convergence regardless of the chosen x_0?

Question 2: Can you find a linear function which, on iteration, appears to give a divergent sequence for *every* initial value?

Question 3: Can you find an example of a linear function whose iteration sequence converges to different limits for different starting values?

Question 4: Can you find a linear function that will give a convergent iteration sequence for one or more initial values and a divergent sequence for other initial values?

Caution: The computer can fool you! Here is an example. Use the computer to try the experiment $(a, b, x_0, n) = (-4, 1, 0.2, 25)$. What do you get? Now, working by hand, try the more modest experiment $(a, b, x_0, n) = (-4, 1, 0.2, 2)$. What do you get? What's going on here? The problem is that while $x_0 = 0.2$ is an exact decimal, computers don't represent numbers decimally. They use binary representations—that is, they express numbers in terms of powers of 2, not powers of 10. The binary representation of 0.2 is *not* exact, and as the computer repeats its calculations for more and more iterations, the round-off error accumulates and becomes more and more noticeable. (This can be a particular problem as you tackle question 4.) Sometimes you can avoid these round-off error problems by choosing values of x_0 which are (positive or negative) powers of 2, like $0.5, 0.25, 0.125$, etc. Later, as we look at the actual proofs, we will not have these problems. The moral here: *computers sometimes compute badly!* Don't put all your stock in them! On the other hand, there is much to be gained by looking at what the computer does tell us, and a wary eye will generally catch the bad output.

Consider also the following questions:

Question 5: What conditions on a and/or b assure that the iteration sequence of $f(x) = ax + b$ converges no matter what the choice of x_0? Guess at this from your examples (and from doing a few more examples with the computer). We'll explore this in detail in what follows.

Question 6: You can think of the set of all linear functions as points in the plane under the correspondence

$$f(x) = ax + b \longleftrightarrow (a, b) \text{ point in } \mathbf{R}^2$$

For example, consider Figure 1.1. With this way of exploring linear

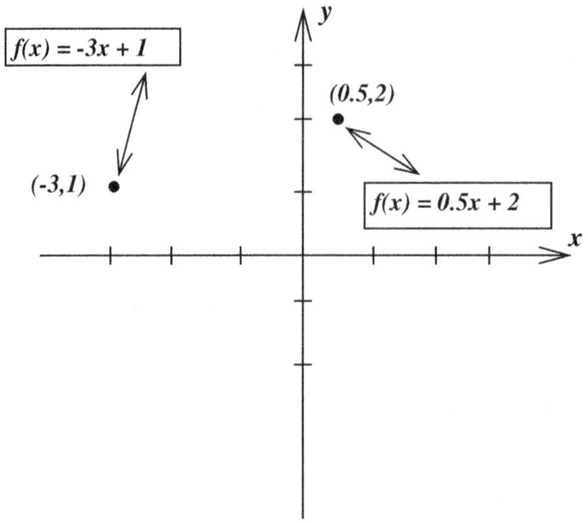

Fig. 1.1: Linear functions corresponding to points

functions under iteration we can think of the (a, b) plane as being cut up into three kinds of pieces:

- *Type (i) points*: those points that give rise to convergent sequences under iteration regardless of the value of x_0,
- *Type (ii) points*: those points corresponding to convergent sequences for some choices of x_0 and divergent sequences for other choices, and

- *Type (iii) points*: points for which there is never convergence.

On a piece of graph paper, draw an (a, b)-plane and, *based on many examples* using your computer program, indicate clearly which points you feel are of Types (i), (ii) and (iii). For instance, the example $f(x) = -3x + 1$ we looked at earlier gave a divergent sequence for most x_0's but converged for one value. Hence this example makes $(-3, 1)$ a Type (ii) point. Similarly, looking at the function $f(x) = 0.5x + 2$, we are convinced that $(0.5, 2)$ is a Type (i) point.

Question 7: For a function $f(x) = ax + b$ always producing a convergent sequence, what is the relationship between the limiting value L and the constants a and b? To explore this question, *fix* a value for a for which you always get convergence. For example, you can try $a = 0.2$ (though you should check this out on the computer for a few values to be sure it is an appropriate value). Now vary b (holding a fixed), noting the limiting values corresponding to each b. On a (b, L) coordinate system, plot the limiting values L against b (all for that fixed a). We show a few points calculated in this way in Figure 1.2.

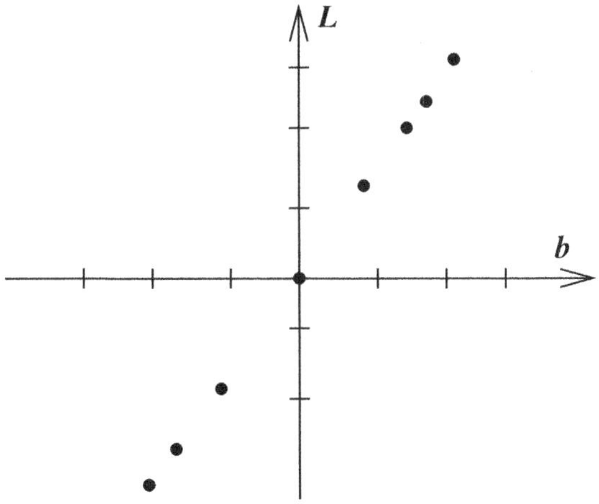

Fig. 1.2: Showing a few values of L against b, for $a = 0.2$

What do you notice? Try this for several fixed a's.

Question 8: If $f(x) = ax + b$ does converge for a starting value of x_0, to what *limit* value L does it converge? What do you suppose determines the limit L; is it a or b or x_0, or a combination of two or all three of these? Make a guess about this on the basis of your experimental evidence. We'll look more at this in the discussion that follows.

Question 9: This question, again, calls for some speculation and some pictures. If a linear function *fails* to give a convergent sequence on iteration, how does it fail? Is there more than one way it could fail to converge? If so, how many ways are there? What does this last question mean? Can you get some experimental data and draw some pictures to show this behavior?

Visualizing iteration. Here is a geometric construction, called a "cobweb" plot, which allows us to visualize the process of iteration. On a piece of graph paper with an (x, y) coordinate system, draw the graph of the line $y = ax + b$ for some choice of a and b. On the same system, draw the graph of $y = x$. Choose a value of x_0 and mark it on the x-axis. Draw a vertical line to the point (x_0, y_0) on the line $y = ax + b$. Now, draw a horizontal line from (x_0, y_0) to the line $y = x$; from the intersection of the horizontal line with $y = x$ draw a vertical to the x-axis. This locates x_1 on the x-axis—why is this true? Repeat the procedure starting with x_1: vertically to (x_1, y_1) on the line $y = ax + b$, horizontally to the intersection with the line $y = x$, and vertically again to the x-axis at x_2, etc. Try some constructions of this same type for linear functions that you know will yield divergent sequences. Try other examples to see what sorts of divergence you can describe.

Question 10: For linear functions which *do* give convergent sequences on iteration, what can you say about how *rapidly* the sequence converges? Begin by deciding what might be a reasonable way to measure speed of convergence. Then, which of the parameters a, b, or x_0 seems to affect the speed?

1.5 Discussion

This section gives you some hints for your write-up of the laboratory and for your analysis of the convergence of the iteration sequences. We have been exploring the sequence $\{x_n\}$ given by

$$x_n = ax_{n-1} + b, \quad n = 1, 2, ...$$

where the starting value x_0 is given. The sequence begins:

$$\begin{aligned}&x_0\\&x_1 = ax_0 + b\\&x_2 = a(ax_0 + b) + b = a^2x_0 + ab + b = a^2x_0 + b(1 + a)\end{aligned}$$

- Write down two additional terms, expanding them algebraically as we have just done (as a multiple of x_0 plus a multiple of b).
- Write your guess for a general term x_n in the same way that you have just done for x_1 through x_4.
- Assume, for the time being, that $a \neq 1$. Recalling that

$$a^{n-1} + a^{n-2} + \cdots + a + 1 = \frac{1 - a^n}{1 - a}$$

observe that this general term can be written in the form

$$x_n = a^n x_0 + \frac{b(1 - a^n)}{1 - a}$$

Now verify, by doing the algebra, that this expression can be rewritten as

$$x_n = a^n \left(x_0 - \frac{b}{1 - a}\right) + \frac{b}{1 - a}$$

- *Mathematical induction* is a method used to give a formal proof of a statement of the form

Condition $P(n)$ holds for every positive integer n.

If you have studied mathematical induction before or if it is covered in your class, use it to *prove* that

$$x_n = a^n \left(x_0 - \frac{b}{1 - a}\right) + \frac{b}{1 - a}$$

is correct for all positive integers n.

- What can you say about $\lim_{n\to\infty} a^n$? In other words, for a particular a, what happens to the value of a^n as n grows larger and larger? How does the value of a affect this limiting behavior? Look at some examples, and try to support your conclusions with a mathematical argument.

- Look carefully at your final expression for x_n. This and your answers to some of the previous questions should suggest two possible conditions that will guarantee convergence. One of the conditions is a property of the function itself (and has nothing to do with x_0). We will call this *Condition I.* The other is a property of both the function and the starting point x_0. This will be *Condition II.* State these conditions, and try to prove that the iteration sequence converges if either of these conditions holds. What is $L = \lim_{n\to\infty} x_n$ under either of these conditions?

- Now return to the (a, b) plane that you considered earlier. With your new (more exact) information, draw a graph in which you shade again values of a and b that yield convergent sequences according to Condition I. Next draw yet another (a, b) plane and shift attention to Condition II. Suppose x_0 is *fixed* at 1. Sketch the set of all points (a, b) that give convergence for this value, and label this set. Do the same for $x_0 = 2$ (on the same graph), for $x_0 = 1/2$, $x_0 = -1$, and a few other values (labeling all the sets). What figures do you get for each x_0?

- If $a = 1$, what condition ensures convergence?

- Finally, are there any values of (a, b) which *never* yield convergence? That is, are there any points that are not included in either of the previous two graphs you drew? If so, prove, by looking at the construction of the sequence x_n, that convergence fails at these points.

1.6 References

Of these three references, the last is more mathematical, the first two are more popular.

- James Gleick, *Chaos*, Viking Press, 1987, (see the chapter on "Life's ups and downs".)

- Ian Stewart, *Does God Play Dice?*, Blackwell Press, 1989 (see chapter 8 of this reference).
- Robert Devaney, *A First Course in Chaotic Dynamical Systems*, Addison-Wesley, 1992.

1.7 Computer programs

1.7.1 *MATLAB/Octave* program

Here is a program that runs in MATLAB or OCTAVE that performs the iteration of a linear function — where we are given the coefficients of the function and the starting value. The program outputs the iteration sequence.

Program: Iterlin

```
function x=iterlin(a, b, x0, N)
%function x=iterlin(a, b, x0, N)
%
%Iterates f(x)=ax+b
%
%Inputs
%  a - slope of function
%  b -intercept of function
%  x0 - initial condition
%  N - number of iterations
%Output
%  x - vector of iterates

x=zeros(1, N); %ititialize x as a vector of zeros
x(1)=x0;       %set initial condition

for i=1:(N-1)
 x(i+1)=a*x(i)+b; %iterate
end
```

This code is a function that will work in *MATLAB* or its free open source alternative *Octave*. It should be saved as **iterlin.m** and placed in a directory accessible to *MATLAB/Octave*. To do this make the directory your current directory or add it to your path in *MATLAB/Octave*.

We can get information about the function by entering the following at the *MATLAB/Octave* prompt in the command window.

```
>> help iterlin
```

To run the function with $(a, b, x_0, n) = (0.5, 2, 5, 10)$ and save the output to x type:

```
>> x=iterlin(0.5, 2, 5, 10)
```

We can plot the result and label the axes using:

```
>> plot(x, 'k-o') %plot
>> xlabel('i')   %label axes
>> ylabel('x_i')
```

While we have not done it for this lab, by altering the program line that calculates $x(i+1) = a * x(i) + b$, you could use a modified form of your program to compute iterates of any reasonable function. (With functions that do not involve parameters a and b, you can simply ignore the inputs for a, and b—giving them any values you wish— or you can modify the first line of the program.)

1.7.2 *Mathematica* programs

Iterations

The following command produces a list of the first n iterations of the function $f(x) = ax + b$ with initial value x_0.

Iteration[a_, b_, x0_, n_]:=NestList[Function[x, $ax+b$], x0, n]

Here is an example: $f(x) = 1.1x - 2.3$, with initial value $x_0 = 0.4$, and 100 iterations. The sequence of iterations seems to diverge.

Iteration[1.1, −2.3, 0.4, 100]

{0.4, -1.86, -4.346, -7.0806, -10.0887, -13.3975, -17.0373, -21.041, -25.4451, -30.2896, -35.6186, -41.4804, -47.9285, -55.0213, -62.8235, -71.4058, -80.8464, -91.231, -102.654, -115.22, -129.041, -144.246, -160.97, -179.367, -199.604, -221.864, -246.351, -273.286, -302.914, -335.506, -371.356, -410.792, -454.171, -501.888, -554.377, -612.115, -675.627, -745.489, -822.338, -906.872, -999.859, -1102.15, -1214.66, -1338.43, -1474.57, -1624.32, -1789.06, -1970.26, -2169.59, -2388.85, -2630.03, -2895.34, -3187.17, -3508.19, -3861.31, -4249.74, -4677.01, -5147.01, -5664.01, -6232.71, -6858.29, -7546.41, -8303.35, -9135.99, -10051.9, -11059.4, -12167.6, -13386.7, -14727.6, -16202.7, -17825.3, -19610.1, -21573.4, -23733.1, -26108.7, -28721.8, -31596.3, -34758.3, -38236.4, -42062.3, -46270.8, -50900.2, -55992.6, -61594.1, -67755.8, -74533.7, -81989.4, -90190.6, -99212., -109135., -120051., -132059., -145267., -159796., -175778., -193358., -212696., -233968., -257367., -283106., -311419.}

Here are the first 80 iterations of the function $f(x) = 0.8x + 2.313$ with initial value $x_0 = 114.7$. The sequence of iterations seems to converge,

Iteration[0.8, 2.313, 114.7, 80]

{114.7, 94.073, 77.5714, 64.3701, 53.8091, 45.3603, 38.6012, 33.194, 28.8682, 25.4075, 22.639, 20.4242, 18.6524, 17.2349, 16.1009, 15.1937, 14.468, 13.8874, 13.4229, 13.0513, 12.7541, 12.5163, 12.326, 12.1738, 12.052, 11.9546, 11.8767, 11.8144, 11.7645, 11.7246, 11.6927, 11.6671, 11.6467, 11.6304, 11.6173, 11.6068, 11.5985, 11.5918, 11.5864, 11.5821, 11.5787, 11.576, 11.5738, 11.572, 11.5706, 11.5695, 11.5686, 11.5679, 11.5673, 11.5668, 11.5665, 11.5662, 11.5659, 11.5658, 11.5656, 11.5655, 11.5654, 11.5653, 11.5652, 11.5652, 11.5652, 11.5651, 11.5651, 11.5651, 11.5651, 11.5651, 11.565, 11.565, 11.565, 11.565, 11.565, 11.565, 11.565, 11.565, 11.565, 11.565, 11.565, 11.565, 11.565, 11.565, 11.565}

Chapter 2

Cyclic Difference Sets

2.1 Introduction

The ideas in this chapter belong to algebra and number theory (and they lead to some interesting geometry). Difference sets have been studied since the 1950's, but recently interest in them has increased because they are associated with "error correcting codes." Error correcting codes aren't intended to keep secrets but rather to enable the receiver of a message to detect and correct transmission errors (the way we can recognize typos in written English and often correct them). For example, information beamed back to earth from Mariner 9 was encoded using mathematics related to difference sets. The *Scientific American* article referred to below also makes use of difference sets, but in this case to create images of distant astronomical objects.

In investigating difference sets, you will:

- Learn something called "modular arithmetic";
- Look for patterns in numerical data;
- Count things in new ways; and
- Discover and perhaps even prove theorems.

In the article "X-Ray Imaging with Coded Masks" (August, 1988, *Scientific American*), the author Gerald K. Skinner uses a "cyclic difference set modulo 15" to create the masks he needs for his X-ray telescope. In this project, you will investigate cyclic difference sets in the cyclic groups that you get when you do arithmetic modulo a positive integer m. Difference sets, in general, might be in other groups.

2.2 Arithmetic modulo 15

To start, let's consider arithmetic "modulo 5," which means we do arithmetic with the integers which are the possible non-negative proper remainders upon division by 5, namely

$$\{0, 1, 2, 3, 4\}.$$

We want to define rules for adding and multiplying these five numbers so that the result is always another of these five numbers. Sometimes we can add or multiply as usual and get answers of the desired kind:

$$0+3=3 \quad 2+1=3 \quad 2\times 2=4$$

But sometimes, the normal result is too large—say $3+4=7$ or $2\times 3=6$. In these cases we replace the sum or product by its positive proper remainder upon division by 5. For example, $7 = 1 \times 5 + 2$, so when we divide 7 by 5 we get remainder 2. Therefore, we say

$$3+4=2 \quad \text{modulo } 5.$$

Similarly, $6 = 1 \times 5 + 1$, so we say

$$2\times 3=1 \quad \text{modulo } 5.$$

Proceeding in this way, we can write out complete addition and multiplication tables modulo 5.

+	0	1	2	3	4
0	0	1	2	3	4
1	1	2	3	4	0
2	2	3	4	0	1
3	3	4	0	1	2
4	4	0	1	2	3

Addition modulo 5

×	0	1	2	3	4
0	0	0	0	0	0
1	0	1	2	3	4
2	0	2	4	1	3
3	0	3	1	4	2
4	0	4	3	2	1

Multiplication modulo 5

Be sure you see how all the entries in the tables are obtained. Notice particularly that

$$2 + 3 = 0 \quad \text{modulo } 5\,,$$

for example, since 5 divided by 5 leaves remainder zero.

Now we do arithmetic modulo 15, which means we do arithmetic with the integers which are the possible non-negative proper remainders upon division by 15, namely

$$\{0, 1, 2, 3, 4, 5, 6, 7, 8, 9, 10, 11, 12, 13, 14\}.$$

As for arithmetic modulo 5, at first we add and multiply these numbers as usual, but if the result is larger than any number on our list of 15 numbers, then we replace it by its positive proper remainder upon division by 15. For example, modulo 15 we have $3 + 6 = 9$, $7 + 11 = 3$ (since $18/15$ leaves remainder 3), $7 \times 2 = 14$, $5 \times 3 = 0$ (since $15/15$ leaves remainder 0), and $5 \times 11 = 10$ (since $55/15$ leaves remainder 10).

Exercise 1: Write out complete addition and multiplication tables modulo 15. (This could be pretty tedious, so you may want to divide up the task among several people—perhaps each person might do 3–5 rows of each table.)

We also need to do subtraction modulo 15, so we need to give meaning to negative numbers. For example, in ordinary arithmetic -3 means the "additive inverse of 3," namely, the number we add to 3 in order to obtain 0. If we add modulo 15, the number we add to 3 in order to obtain 0 is 12; that is, $3 + 12 = 0$ (since $15/15$ leaves remainder 0), so

$$-3 = 12 \quad \text{modulo } 15\,.$$

Exercise 2: Find the additive inverse of each of $0, 1, 2, \ldots, 14$ modulo 15.

We can give meaning to $x - y$ modulo 15 by

$$x - y = x + (-y) \quad \text{modulo } 15.$$

Now we can verify that the set $D = \{1, 2, 3, 5, 6, 9, 11\}$ of the *Scientific American* article is a *cyclic difference set* modulo 15: for each of the 14 possible non-zero values of a in $\{0, 1, 2, \ldots, 14\}$, there are the *same* number of pairs of elements in D whose difference is a (3 pairs in this case):

$a = 1$	$a = 2$	$a = 3$	$a = 4$	$a = 5$	$a = 6$	$a = 7$
$2 - 1 = 1$	$3 - 1 = 2$	$5 - 2 = 3$	$5 - 1 = 4$	$6 - 1 = 5$	$9 - 3 = 6$	$9 - 2 = 7$
$3 - 2 = 1$	$5 - 3 = 2$	$6 - 3 = 3$	$6 - 2 = 4$	$11 - 6 = 5$	$11 - 5 = 6$	$3 - 11 = 7$
$6 - 5 = 1$	$11 - 9 = 2$	$9 - 6 = 3$	$9 - 5 = 4$	$1 - 11 = 5$	$2 - 11 = 6$	$1 - 9 = 7$

$a = 14$	$a = 13$	$a = 12$	$a = 11$	$a = 10$	$a = 9$	$a = 8$
						$2 - 9 = 8$
						$11 - 3 = 8$
						$9 - 1 = 8$

Exercise 3: Complete the table to check that there are exactly 3 pairs of elements in D whose difference is a for $a = 9, 10, 11, 12, 13, 14$. Why do you suppose the values of a listed in the partially completed check table above are arranged the way they are?

2.3 Cyclic difference sets modulo m

Let's choose a positive integer m, and we'll write $\mathbb{Z}/m\mathbb{Z}$ to represent the set of non-negative proper remainders upon division by m,

$$\mathbb{Z}/m\mathbb{Z} = \{0, 1, 2, \ldots, m - 1\}.$$

We call $\mathbb{Z}/m\mathbb{Z}$ the set of *integers modulo m*. We add and multiply the elements of $\mathbb{Z}/m\mathbb{Z}$ modulo m.

Exercise 4: Try some examples of addition and multiplication modulo 6. Now try some modulo 7.

Exercise 5: If x is in $\mathbb{Z}/m\mathbb{Z}$, explain why $m - x$ is as well. Explain why $m - x$ is the additive inverse of x modulo m.

We write

$$a \equiv b \pmod{m}$$

to indicate that a and b leave the same remainder upon division by m, and we say that a and b are **congruent modulo** m.

Exercise 6: If $a \equiv b \pmod{m}$, what can you say about the difference $a-b$? Conversely, for two integers a and b, what condition on $a-b$ guarantees that $a \equiv b \pmod{m}$?

Exercise 7: Explain why *every* integer is congruent modulo m to exactly one element of $\mathbb{Z}/m\mathbb{Z}$.

Exercise 8: Explain why working modulo m assures that the sum or product of two elements of $\mathbb{Z}/m\mathbb{Z}$ is again in $\mathbb{Z}/m\mathbb{Z}$.

A subset D of $\mathbb{Z}/m\mathbb{Z}$ is a **cyclic difference set modulo m** provided that the set of all non-zero differences of elements of D represents each non-zero element of $\mathbb{Z}/m\mathbb{Z}$ the *same* number of times. (Notice that there are *two* requirements here: every non-zero element of $\mathbb{Z}/m\mathbb{Z}$ occurs as a difference of elements of D, and each non-zero value appears as a difference exactly the same number of times.) A cyclic difference set D has three parameters:

$$\begin{aligned} m &= \text{the modulus (and number of elements in } \mathbb{Z}/m\mathbb{Z}) \\ k &= \text{the number of elements in } D \\ \lambda &= \text{the number of pairs of elements of } D \text{ giving} \\ &\quad \text{each non-zero difference modulo } m \end{aligned}$$

In the example above, $m = 15$, $k = 7$, and $\lambda = 3$.

Many methods are known for constructing cyclic difference sets modulo m, but it is an unsolved problem to describe *all* cyclic difference sets. We are going to use the computer to explore one method in detail. (The one used to construct the example in the *Scientific American* article is quite different. It uses some interesting ideas from geometry.)

It's easiest to describe the method we will explore by using it in a specific case. We will do our example with $m = 7$, and we will find that $k = 3$ and $\lambda = 1$. Since $m = 7$, we are working with the integers modulo 7: $\mathbb{Z}/7\mathbb{Z} = \{0, 1, 2, 3, 4, 5, 6\}$. The method requires that we square each of the non-zero elements of $\mathbb{Z}/7\mathbb{Z}$: $1 \times 1 = 1$, $2 \times 2 = 4$, $3 \times 3 = 2$, $4 \times 4 = 2$, $5 \times 5 = 4$, $6 \times 6 = 1$.

Exercise 9: Why does $1 \times 1 = 6 \times 6$? Why does $2 \times 2 = 5 \times 5$? Why does $3 \times 3 = 4 \times 4$?

We take for the set D the distinct non-zero squares of elements of $\mathbb{Z}/7\mathbb{Z}$:

$$D = \{1, 2, 4\}.$$

Clearly $k = 3$, but is this D really a cyclic difference set with exactly 1 pair of elements in D whose difference is a for $a = 1, 2, 3, 4, 5, 6$?

Exercise 10: Check that this D is a cyclic difference set modulo 7 with $\lambda = 1$.

Unfortunately, this method does not work for every choice of modulus m.

Exercise 11: (a) Carry out this procedure with $m = 5$ and check whether the resulting set of distinct non-zero squares of elements of $\mathbb{Z}/5\mathbb{Z}$ is or is not a cyclic difference set. (b) Now try $m = 6$.
(c) What about $m = 15$?

2.4 Questions to explore

Now we come to the main questions for this investigation. Choose a modulus m and form the set D of the non-zero squares modulo m. When m is relatively small, this is easy to do by hand. For larger values of m it is helpful to use a computer. Here is the logical structure of a computer program called SQUARES that creates the set of distinct non-zero squares modulo m. Be aware that for extremely large values of m the computer may crash due to memory limitations.

Program outline: Squares

```
Input: the modulus m
Output: the distinct non-zero squares modulo m

S := [1,2,...,m-1]  %create array of the integers modulo m
S := S.^2           %square each entry of S
S := mod(S,m)       %replace entries of S by their values mod m
US := unique(S)     %find the unique elements of the array S
                    %and assign to US
OUTPUT US
```

Be sure you can explain why this program does what it should.

Question 1: For which values of m is the number k of distinct elements in D equal to $(m-1)/2$? (Can m be even in this case?)

NOTE: For the next two questions we will work *only* with values of m for which D has exactly $(m-1)/2$ elements.

Now we will investigate in which of these cases D is actually a difference set. Since calculating all possible differences of elements of D by hand is very tedious, we need a computer program. Below is pseudocode for the program DIFFERENCES. Be sure you can explain why this program does what it should.

Question 2: Find at least 10 values of m for which D is a cyclic difference set. Can you characterize these "good" values of m?

Question 3: For "good" values of m (in the sense of Question 2), let k be the number of elements in D, and let λ be the number of pairs of elements in D giving each non-zero difference modulo m. Can you express the parameters k and λ as functions of m?

Program outline: Differences

```
Input:  the modulus m
Output:  the distinct non-zero squares, their non-zero
         differences, and the number of repetitions

sq := squares(m)    %uses squares program to get list
                    %of nonzero squares
N := length(m)      %number of nonzero squares
INITIALIZE freq as an m-1 dimensional vector of zeros
FOR i = 1 to N
  FOR j = 1 tp N
     diffij=mod(sq(i)-sq(j),m) %compute difference of ith
                               %and jth squares mod m
     IF diffij > 0 THEN
        freq(diffij)=freq(diffij)+1   %increase count of
                                      %occurance of diff by 1
     END IF
  END FOR
END FOR
diff=[1,2,...,m-1]      %a vector of all possible positive
                        %integers mod m
diff=diff(freq>0)       %only keep differences which had
                        %nonzero frequency
freq=freq(freq>0)       %only keep nonzero frequencys

OUTPUT sq diff freq
```

2.5 Discussion

The goal of this section is to provide analytic support for at least some of your empirical observations. The next five questions are meant to help you *prove* that the pattern you see in Question 1 always holds.

Question 4: Suppose x is in $\mathbb{Z}/m\mathbb{Z}$ and y is the additive inverse of x. Can you show that $x^2 \equiv y^2 \pmod{m}$?

Question 5: If $x \neq 0$ is in $\mathbb{Z}/m\mathbb{Z}$, can x equal its additive inverse modulo m? Under what circumstances?

Question 6: What do the answers to Questions 4 and 5 say about the *maximum* number of distinct non-zero squares modulo m when m is odd? when m is even?

NOTE: For the remaining questions in this section, assume that m is odd.

Question 7: Suppose m is a perfect square, $m = a^2$ for some integer a. Can you show that in this case the number of distinct non-zero squares modulo m must be *less* than $(m-1)/2$?

Question 8: What can you say about the case when $m = ab$ for $0 < b < a < m$? [Hint: compare $(a+b)^2$ and $(a-b)^2$ modulo m. Can $a+b$ and $a-b$ be congruent modulo m? Can they be additive inverses modulo m?]

Question 9: Can you prove that the formulas for k and λ in terms of m that you found in Question 3 are correct? [Hint: There are k elements in D. In how many ways can you form a difference $x - y$ with x and y distinct elements of D? The number of such differences depends on k. If D is a difference set, the number of such differences must also be related to λ and m.]

Question 10: Try to prove that your characterization of the "good" values of m (in the sense of Question 2) is correct. [It is probably too hard to prove that if m has the right form, then D is a cyclic difference set; but try to prove that if D is a cyclic difference set, then m has a particular form.]

2.6 Computer programs

2.6.1 *MATLAB/Octave* programs

Program: Squares

```
function us=squares(m)
%function us=squares(m)
%
%Finds the unique squares mod m
%
%Inputs
% m - integer to compute squares modulo m
%Outputs
% us - vector of unique squares modulo m

s=(1:(m-1)).^2;
s=mod(s,m);
us=unique(s);
```

Program: Differences

```
function [sq,diff,freq]=differences(m)
%function [sq,diff,freq]=differences(m)
%
%Finds the unique squares mod m, the list of unique non-zero
%differences of squares modulo m and their frequency of occurance.
%
%Inputs
% m - integer for computation modulo m
%Outputs
% sq - vector of unique squares modulo m
```

```
% diff - vector of non-zero differences of squares modulo m
% freq - vector of frequencies of occurance of the differences.

sq=squares(m);    %find unique squares mod m using our squares function
N=length(sq);     %number of unique squares mod m
freq=zeros(1,m-1); %initialize vector of frequencies to array of zeros

for i=1:N
   for j=1:N
      diffij=mod(sq(i)-sq(j),m); %difference of ith and jth squares mod m
      if diffij > 0
         freq(diffij)=freq(diffij)+1; %increase count of occurance
                                      %of diff by 1
      end %if
   end %for
end %for

diff=1:m-1;             %all possible positive integers mod m
diff=diff(freq>0);      %only keep differences with nonzero frequency
freq=freq(freq>0);      %only keep nonzero frequencies
```

This function has multiple outputs and should be called as follows, for example with $m = 7$.

```
>> [sq,diff,freq]=differences(7)
```

The function produces three separate vectors: sq, $diff$, and $freq$. We may want to display the differences and frequencies aligned in a single table. To accomplish this we can use the following command:

```
>> [diff;freq]
```

2.6.2 *Mathematica* programs

Basics

Note: Be sure to run the following series of commands first, before anything else!

```
Squares[m_]:=Drop[Union[Mod[#^2, m]&/@Range[m]], 1];

GenDiff[D_, m_]:=Mod[Outer[Subtract, D, D], m];
VerticalLine[n_]:=Table["|", {n}];
HorizontalLine[n_]:=Table["–", {n}];
PadMatrix[corner_, top_, left_, body_]:=
    Transpose[Prepend[Transpose[Prepend[body, top]], Prepend[left, corner]]];
GenDiffTable[D_, m_]:=MatrixForm[PadMatrix["-", Prepend[D, "|"], Prepend[D, "–"],
    PadMatrix["+", HorizontalLine[Length[D]], VerticalLine[Length[D]], GenDiff[D, m]]]];
DiffTable[m_]:=GenDiffTable[Squares[m], m];

RawFreq[S_, n_]:=Count[S, #]&/@Range[n];
RawFreqTable[S_, n_]:={Range[n], RawFreq[S, n]};
FreqTable[S_, n_]:=MatrixForm[RawFreqTable[S, n]];
FreqLongTable[S_, n_]:=MatrixForm[Transpose[RawFreqTable[S, n]]];
GenDiffFreqTable[D_, m_]:=FreqTable[Flatten[GenDiff[D, m]], m - 1];
GenDiffFreqLongTable[D_, m_]:=FreqLongTable[Flatten[GenDiff[D, m]], m - 1];
DiffFreqTable[m_]:=GenDiffFreqTable[Squares[m], m];
DiffFreqLongTable[m_]:=GenDiffFreqLongTable[Squares[m], m];

GenDiffFreq[D_, m_]:=Union[RawFreq[Flatten[GenDiff[D, m]], m - 1]];
DiffFreq[m_]:=GenDiffFreq[Squares[m], m];
```

Examples with squares

m=7

First, we find the (non-zero) squares (mod 7), to be used as the our possible cyclic difference set. Note that $1^2 = 1$, $2^2 = 4$, $3^2 = 9 \equiv 2 \pmod 7$, $4^2 = 16 \equiv 2 \pmod 7$, $5^2 = 25 \equiv 4 \pmod 7$, and $6^2 = 36 \equiv 1 \pmod 7$.

Squares[7]

$\{1, 2, 4\}$

Next, we compute all the possible differences among this set. Note , for instance, that the entry in the middle row, right column is $2 - 4 = -2 \equiv 5 \pmod 7$.

DiffTable[7]

$$\begin{pmatrix} - & | & 1 & 2 & 4 \\ - & + & - & - & - \\ 1 & | & 0 & 6 & 4 \\ 2 & | & 1 & 0 & 5 \\ 4 & | & 3 & 2 & 0 \end{pmatrix}$$

Something nice about the set {1,2,4} (mod 7) is that in the above difference table, every non-zero entry occurs the same number of times! It's easy to see this in this small case from the above table, but as the numbers get larger, it will be nice to have a program that checks the number of occurrences of each entry. The below program verifies that every number from 1 through 6 occurs exactly once.

DiffFreqTable[7]

$$\begin{pmatrix} 1 & 2 & 3 & 4 & 5 & 6 \\ 1 & 1 & 1 & 1 & 1 & 1 \end{pmatrix}$$

When the numbers get larger, it may be handy to list this table vertically instead of horizontally:

DiffFreqLongTable[7]

$$\begin{pmatrix} 1 & 1 \\ 2 & 1 \\ 3 & 1 \\ 4 & 1 \\ 5 & 1 \\ 6 & 1 \end{pmatrix}$$

Another quick check, handy as the numbers get larger, is to simply list all the frequencies in the above table. If there is only one frequency, we have a cyclic difference set, and λ is simply equal to this unique frequency. Below, this shows that λ=1 here.

DiffFreq[7]

{1}

Also, we can compute k, the number of elements in the cyclic difference set, as below.

Length[Squares[7]]

3

Chapter 3

The Euclidean Algorithm

3.1 Introduction

In this chapter we consider an ancient and ingenious observation about the integers, the Euclidean algorithm. You will

- Learn the Euclidean algorithm;
- Investigate the speed of the algorithm;
- Use the algorithm to investigate properties of the integers; and
- Use the algorithm to solve linear Diophantine equations.

Incidentally, the word "algorithm" refers to a fixed set of directions which gives a procedure for accomplishing some task. In grade school, we learned many algorithms. Examples include the algorithms for multiplying integers and decimal numbers and for dividing one integer into another. The Euclidean algorithm provides a procedure for finding the largest integer that divides (evenly) two given integers. It is one of the oldest, and most important algorithms in mathematics: it generalizes to many other number systems and algebras in which there are notions of multiplication and addition and something which generalizes the ordinary concept of "less than" for integers.

3.2 The algorithm

When you first learned about fractions, what you now call rational numbers, you encountered the notion of putting a fraction in "lowest terms." Your teacher may have insisted that you not write $6/15$, for example, but rather that you notice that 3 is a factor in both the numerator and denominator, which can be canceled, so that you should write $2/5$. There are no other integer factors common to numerator and denominator, so the fraction is now in lowest terms. Perhaps, though, your teacher was more permissive. You could be forgiven for writing $5117/6923$, even though that fraction is not in lowest terms. The numbers 5117 and 6923 both contain 301 as a factor, and accordingly you should cancel it and obtain $17/23$. How would you know that though?

The problem of reducing a fraction a/b to lowest terms is the task, given two integers a and b, of finding the greatest integer which divides both of them, the greatest common divisor (gcd), also sometimes called highest common factor (hcf). We will write, for example

$$\begin{aligned} \gcd(6, 15) &= 3 \\ \gcd(5117, 6923) &= 301 \end{aligned}$$

Before going any further, you might wish to try computing a few gcd's.

Exercise 1: Compute $\gcd(81, 42)$, $\gcd(72, 95)$ and $\gcd(1336517, 1304051)$.

The problem of computing $\gcd(a, b)$ has a very elegant solution, the **Euclidean algorithm**, known from ancient times. We will illustrate it in computing $\gcd(5117, 6923) = 301$. We start with the two integers 5117 and 6923, and divide the smaller into the larger, getting the quotient 1 and the remainder 1806. Now we discard 6923, the larger of the two integers we started with, promote 5117 to the position of larger integer, and take the remainder 1806 as the new smaller integer. Now just repeat: divide 1806 into 5117, getting the quotient 2 and the remainder 1505. Discard the larger 5117, promote

1806 to be the larger integer and take 1505 as the smaller. Divide 1505 into 1806, getting quotient 1 and remainder 301 (aha!). Discard 1806, divide 301 into 1505, getting quotient 5 with remainder 0. Since 301 divides in evenly, it is the gcd! Now let's look at the sequence of calculations:

$$\begin{aligned} 6923 &= 5117(1) + 1806 \\ 5117 &= 1806(2) + 1505 \\ 1806 &= 1505(1) + 301 \\ 1505 &= 301(5) + 0. \end{aligned}$$

Therefore, gcd(6923, 5117) = 301

Why does it work? The explanation goes back to the meaning of 'quotient' q and 'remainder' r. If a and b are integers, with $b \neq 0$, then by division we find q and r with $0 \leq r < |b|$ such that

$$a = qb + r\,.$$

We can make a simple geometric argument to demonstrate this important fact about the integers and, in particular, to show how the bounds on the remainder are obtained. On the real line, in the case that b is positive mark off all integer multiples of b as in Figure 3.1. In the case that b is negative mark them off as in Figure 3.2.

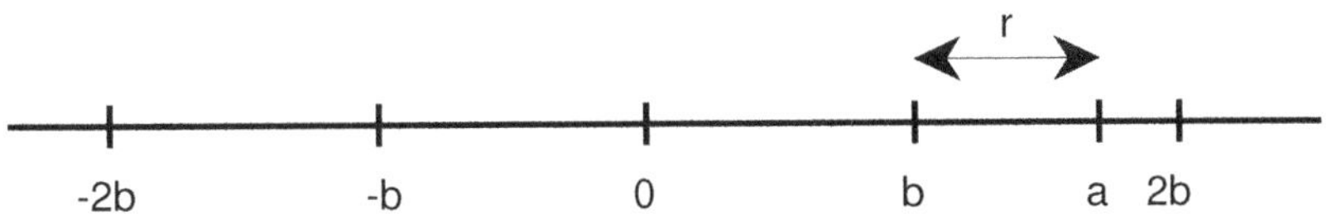

Fig. 3.1: b positive and a positive

Now clearly the dividend a is itself a multiple of b, or it must lie between two multiples of b, say between qb and $(q+1)b$. In the first case, $r = 0$. In the second case, we can choose the multiple qb to the left of a, so $qb < a$. Then $r = a - qb$ is positive and

$$0 < r < |b|.$$

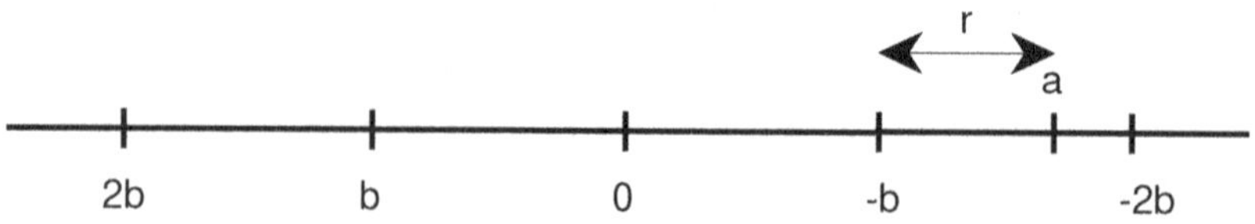

Fig. 3.2: b negative and a positive

(Note that we could have chosen the multiple of b which lies closer to a. In that case, we may not have a positive remainder; the following bound will hold instead: $0 < |r| \leq |b|/2$.)

Now, since

$$r = a - qb,$$

any integer which divides both a and b divides the right hand side, and therefore divides the integer r. In particular $\gcd(a, b)$ divides both b and r, so $\gcd(a, b) \leq \gcd(b, r)$. But since $a = qb + r$, any divisor of both b and r, including $\gcd(b, r)$, is a divisor of a, so $\gcd(b, r) \leq \gcd(a, b)$ (same inequality, but going the other way), and we therefore conclude

$$\gcd(a, b) = \gcd(b, r).$$

This is recognizable as the key step in the Euclidean algorithm: b is promoted to the place of a, and r takes the place of b. Note that the algorithm is unaffected after the first step or two by the signs or relative sizes of a and b; that is, even if a or b or both are negative or if $b < a$, we soon arrive at a step in the process after which both r_i and r_{i+1} are positive and $0 \leq r_{i+1} < r_i$. For example, suppose that $a, b < 0$ and that $|a| < |b|$. We then obtain the diagram shown in Figure 3.3.

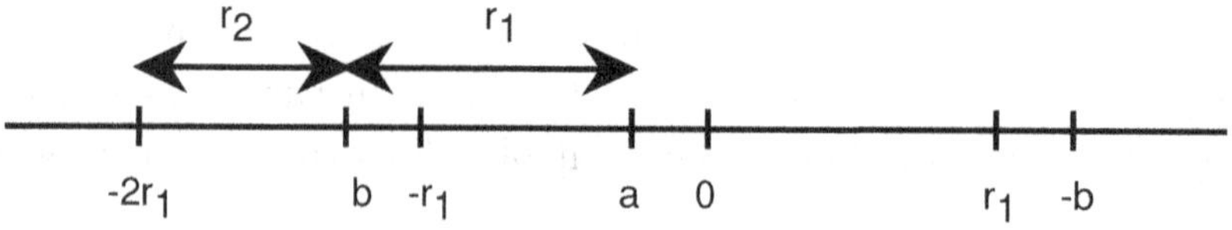

Fig. 3.3: $a, b < 0$; $|a| < |b|$

The corresponding calculations are

$$
\begin{aligned}
a &= b(1) + r_1, \qquad (0 < r_1 < |b|) \\
b &= r_1(q) + r_2 = -2r_1 + r_2 \qquad (0 < r_2 < r_1) \\
r_1 &= r_2(q_2) + r_3
\end{aligned}
$$

and so on. Since $b < a$ and $r < b$, both integers have strictly decreased. If $r = 0$, then b was the gcd. If $r > 0$, we have the same type of problem, but with smaller integers, so we repeat the process. In the above example it looked like this:

$$
\begin{aligned}
\gcd(5117, 6923) &= \gcd(1806, 5117) \\
&= \gcd(1505, 1806) \\
&= \gcd(301, 1505) \\
&= \gcd(0, 301) \\
&= 301
\end{aligned}
$$

The process is a kind of iteration, but unlike the iteration in chapter 1, we do not need to worry about the issue of convergence: the Euclidean algorithm finishes in a *finite* number of steps. It is not an infinite process. The reason is that in each step the remainder r in the division is *strictly less* than in the step before. You can see this in the example: the remainders were 1806, 1505, 301, and 0, and it is clear from the definition of quotient and remainder, in particular the condition $r < |b|$, that the sequence of remainders must decrease like this. Since the sequence of integer remainders is strictly decreasing, but bounded below by zero, the process must terminate in a finite number of steps. It is clear in fact that the number of steps cannot be bigger than $|b|$, because (one might think) even in the worst case the sequence of remainders would be $b-1$, $b-2$, ... , 1, 0. Actually this sequence cannot occur. The Euclidean algorithm works *much* faster than that!

Exercise 2. Suppose that $|a| < |b|$. Show that the Euclidean algorithm proceeds as described when

1. $a > 0$ and $b > 0$,
2. $a > 0$ and $b < 0$,
3. $a < 0$ and $b > 0$.

3.3 Questions and discussion

Below is a description of the Euclidean algorithm in pseudocode which prints out the result of each step of the algorithm and counts the number of steps needed to finish.

Program outline: Euclid1

```
Input:  positive integers a and b, a > b.
Output: steps of the Euclidean algorithm, giving gcd(a,b)

as := a
bs := b
% as and bs are initially 1x1 dynamic arrays which will
% increase in length as the program progresses.

i := 1
WHILE bs(i) > 0
as(i+1) := bs(i)
bs(i+1) := the remainder of as(i)/bs(i)
i := i + 1
END WHILE

OUTPUT as, bs

%The final value of the as array will be the gcd(a,b)
%the final value of the array b will be 0.
```

The number of steps to complete the algorithm is simply the length of the array as minus one to account for the initial values of a and b. The arrays as and bs give the sequence of pairs of numbers with equal gcd.

$$\gcd(a(1), b(1)) = \gcd(a(2), b(2)) = \gcd(a(3), b(3)) \dots$$

The pessimistic bound at the end of the introduction suggested that if a and

b are around 10,000, then it might take around 10,000 steps to compute the gcd.

Question 1: Investigate this in a systematic way by running the program EUCLID1 with various pairs (a, b), taking $1 < b < a < 100$, then $100 < b < a < 1000$, then $1000 < b < a < 10000$, etc. (Careful: your computer language may not handle integers larger than 32767 correctly.) Do bigger numbers take more steps? How does the dependence go? What is the worst case, in the sense of taking the most steps? You will notice the program runs very fast, even when the inputs are large. The number of steps is never very large—is this just lucky, or is there really a much stricter upper bound on the number of steps than the number b? Look at how the sequence of remainders behaves, and try to construct a better bound.

Are you tired of thinking up integer pairs? Let the computer do this for you by using a random number generator to choose pairs (a, b) at random. We create a new program EUCLID2 which chooses at random a pair of integers (a, b), each between 1 and some preassigned number N, uses the Euclidean algorithm to compute their gcd, showing all intermediate steps and counting the number of times it takes the algorithm to terminate.

Program outline: Euclid2

```
Input: a positive integer N
Output: a randomly chosen pair of positive integers a and b,
        0 < b <= a < N, and the steps of the Euclidean
        algorithm, leading to gcd(a,b).

a := a random integer between 1 and N
b := a random integer between 1 and N
RUN EUCLID1
numsteps := length(as) - 1  %length of as vector from
                            %EUCLID1 minus 1
gcd = final entry of as

OUTPUT gcd, numsteps, a b
```

Computer languages have different ways of choosing random numbers between specified values. For example, in *MATLAB/Octave*, each time you enter *rand*, the program picks a random (decimal) number between 0 and 1. Thus, to choose a random number a between 1 and N, you would enter a line such as "`a := integer part of (1 + rand*N)`." As another example, to choose a number a at random between integers A and B you would enter a line like '`a := integer part of (A + rand*(B-A))`."

Now you have a quick way to generate random pairs of integers between 1 and N, and the pairs are not biased (the way your own choices probably were), so you can do meaningful statistics. Try the program EUCLID3 which follows.

Program outline: Euclid3

```
Input:  two positive integers N, M
!corresponding to choosing M pairs of integers
!between 1 and N at random
Output: number of times in M trials that the Euclidean
algorithm required 1 step, 2 steps,..., 20 steps
to terminate

INITIALIZE freq as a vector of 20 zeros
%these will seve as counters
FOR j = 1 TO M
RUN EUCLID2 with input N (but don't print the output)
IF the output numsteps  < 21, THEN
freq(numsteps) := freq(numsteps) +1
END IF
END FOR

OUTPUT freq
```

The array entries `freq(numsteps)` keep track of how many times in M random trials the Euclidean algorithm required 1 step, 2 steps, etc. You may not see the worst case—after all, there are 10^8 integer pairs if $N = 10000$, and if, for example, $M = 1000$, you are only looking at a minuscule fraction of them. But you will see typical cases. Try changing N to see how the performance of the algorithm changes.

Question 2: Re-do question 1 using the program EUCLID3 with, say, 1000 randomly chosen pairs (a, b), taking $10 < b < a < 100$, then $100 < b < a < 1000$, then $1000 < b < a < 10000$, etc. Plot the number of times the Euclidean algorithm requires k steps against different values of k, $1 \leq k \leq 20$. How does the shape of this plot change for $10^n < a, b < 10^{n+1}$, as n changes from 1 to 7? Do bigger numbers take more steps on average? How does the dependence go?

Another interesting thing to keep track of is the gcd's themselves. It is clear that any positive integer m is the gcd of *some* integer pair; for example, $\gcd(2m, m) = m$. But some numbers seem to turn up as gcd's more often than others. You can keep track of this by making a minor change in the program above, so that it counts the occurrences of gcd values instead of the number of occurrences of stopping values. Create EUCLID4 by replacing the lines in EUCLID3

```
IF the output  numsteps  < 21, THEN
   freq(numsteps) := freq(numsteps) +1
END IF
```

by

```
IF the output  gcd  < 21, THEN
   freq(gcd) := freq(gcd) +1
END IF
```

Recall that the value "numsteps" is the number of times it took for the Euclidean algorithm to terminate for that pair.

You can also try this with different N's, and if you want even better statistics, you could run it with M random pairs, $M > 1000$. Is there a pattern here?

Question 3: Plot the relative frequency of occurrence of 1,2,3,... as $\gcd(a, b)$ against 1,2, 3, ... when a and b are chosen in the interval

$$10^n < b < a < 10^{n+1}.$$

Do this for several values of n. How do these plots depend on n?

If $\gcd(a,b) = 1$, then the greatest integer which divides both a and b is 1, but of course 1 divides every integer, so in this case a and b have *no* non-trivial factors in common. They are said to be **relatively prime** (or **coprime**). In the language of grade school, a/b is already in lowest terms.

Question 4: What is the probability that integers a and b chosen randomly are relatively prime? You already have a rough answer to this question if you did Question 4: it is just the fraction of the time that the integers a and b are found to have gcd=1. This happens more than half the time. A clever argument says that the probability is actually $6/\pi^2 \approx 0.6079$. How does this compare with your experimental determination?

Question 5: Can you construct an argument to show that the fraction of pairs of relatively prime integers is $6/\pi^2$? Look carefully at the distribution of gcd values and ask yourself how much more probable gcd=1 is than gcd=2 (that is, find the ratio of the number of pairs with gcd =2 to gcd = 1). Ask the same question for gcd=3 (that is, find a ratio again). Why is your answer valid? It also helps to notice that the number of integer pairs with both a and b *even* numbers is 1/4 of all integer pairs. (Caution: this plausible statement is very imprecise!) The case a and b both even is not the same thing as gcd(a,b)=2, of course, since for many such pairs the gcd would be an even number larger than 2, but it is suggestive. Similarly, the number of integer pairs with a and b both multiples of 3 is 1/9 of all integer pairs, etc. Finally you would need to know a famous result of Euler:

$$\sum_{j=1}^{\infty} \frac{1}{j^2} = 1 + \frac{1}{4} + \frac{1}{9} + \ldots = \frac{\pi^2}{6}$$

The **Fibonacci sequence** $\{n_i\}$ is defined by

$$n_1 = 1, \quad n_2 = 1, \qquad n_{j+1} = n_j + n_{j-1}, \text{ for } j \geq 2$$

This sequence has many fascinating properties. The next question invites you to use the Euclidean algorithm on Fibonacci numbers.

Question 6: What is $\gcd(n_j, n_{j+1})$? (i.e., look at adjacent Fibonacci numbers). How many steps does the algorithm take? Do Fibonacci pairs behave like your randomly chosen pairs in the previous section? What about $\gcd(n_j, n_{j+2})$? Is there a pattern here? Try $\gcd(n_j, n_{j+3})$.

Your observations should suggest propositions you can try to prove using the definition of Fibonacci numbers

$$n_{j+1} = n_j + n_{j-1}$$

and the basic step

$$\gcd(a, b) = \gcd(b, r)$$

in the Euclidean algorithm. In this connection you will find yourself asking, if you know $\gcd(a, b)$, does this help to find $\gcd(a, mb)$, where m is an integer? This is a question about multiplication of integers in general.

3.4 Linear Diophantine equations

Let a, b,and c be integers. We can interpret the linear equation

$$ax + by = c$$

as the equation of a line in the x-y plane. If someone were to ask us to find solutions (x, y), we might be justifiably puzzled—that's what that line is! There are infinitely many solutions. But if it were further required that x and y also be integers, then the situation is not so clear. This is asking for just those points on the line which happen to have integer coordinates. Geometrically, these are points where the line $ax + by = c$ hits the *integer lattice*, the points of the plane with integer coordinates. But does this happen at all? An equation which is required to have integer solutions is called a **Diophantine equation**.

If the Diophantine equation $ax + by = c$ has solutions, then $\gcd(a, b)$ divides the left side, so $\gcd(a, b)$ must divide c. This is clearly a necessary condition for a solution to exist. Since it might not be satisfied, we see that some Diophantine equations do *not* have solutions. On the other hand, if $\gcd(a, b)$ divides c, then this Diophantine equation *does* have solutions: the condition is sufficient. In this case the Euclidean algorithm produces a solution (x, y). More precisely, it produces a solution (x', y') to

$$ax' + by' = \gcd(a, b)$$

and then

$$(x, y) = (c'x', c'y') \tag{3.1}$$

solves the original problem, where $c' = c/\gcd(a, b)$.

To see that this is true requires that we look at the algorithm in more detail. The Euclidean algorithm produces a sequence of remainders, each of the form $ax + by$:

$$\begin{aligned}
r_1 &= a - bq_1 = a \cdot 1 + b(-q_1) \\
r_2 &= b - r_1 q_2 = b - (a - bq_1)q_2 = a(-q_2) + b(1 + q_1 q_2) \\
r_3 &= r_1 - r_2 q_3 = a(1 + q_2 q_3) + b(-q_1 - q_3(1 + q_1 q_2)) \\
&\vdots \\
0 &= r_{n-2} - r_{n-1} q_n
\end{aligned}$$

In the nth (last) line, we see $r_{n-1} = \gcd(a, b)$, since r_{n-1} divides evenly into r_{n-2}, and r_{n-1} is of the form $ax + by$ because all the remainders are. We have shown this for r_1, r_2, and r_3 explicitly in the first three lines. We can also extend the notation, preserving the pattern, so that

$$\begin{aligned}
r_{-1} &= a = 1 \cdot a + 0 \cdot b \\
r_0 &= b = 0 \cdot a + 1 \cdot b
\end{aligned}$$

Now we just have to keep track of the coefficients of a and b as we use the Euclidean algorithm, and note the result in line $n - 1$. If we label the coefficients in the obvious way

$$r_j = ax_j + by_j$$

then for any $k > 0$

$$r_k = r_{k-2} - r_{k-1} q_k = a(x_{k-2} - x_{k-1} q_k) + b(y_{k-2} - y_{k-1} q_k)$$

which is a recursion relation for the coefficients:

$$\begin{aligned}
x_k &= x_{k-2} - x_{k-1} q_k \\
y_k &= y_{k-2} - y_{k-1} q_k
\end{aligned}$$

Starting with $x_{-1} = 1$, $y_{-1} = 0$, $x_0 = 0$, $y_0 = 1$ from the equations for r_{-1} and r_0, we can use this recursion relation $n-1$ times to find $(x, y) = (x_{n-1}, y_{n-1})$. This sort of thing is best done by a computer, of course. Below is pseudocode for the program EUCLID5.

Program outline: Euclid5

```
Input: integers a and b, with a > b > 0.
Output: integer solutions to ax + by = gcd(a,b)

a1 := a, b1 := b
xold := 1, yold := 0
x := 0, y := 1
c := 1.
WHILE c ~= 0  % The symbols ~= means not equal to
    Let q := int(a/b), c := a - q*b %quotient and remainder
    IF c = 0 THEN
        print GCD = x*a1 + y*b1
    END IF
    xnew := xold - q*x, ynew := yold - q*y
    xold := x, yold := y
    x := xnew, y := ynew, a := b, b := c
END WHILE
```

Exercise 3: Try EUCLID5 on some examples.

The Euclidean algorithm gives a particular solution to the Diophantine equation $ax + by = c$, if there are any solutions at all. In fact, if there is one, there is an infinite number, regularly spaced along the line. Two distinct solutions would mean

$$ax + by = ax' + by' = c$$

so that

$$\frac{a}{\gcd(a,b)}(x - x') = \frac{b}{\gcd(a,b)}(y' - y)$$

Since the first factors on each side do not have any factors in common (these have been divided out), it must be that they divide $x - x'$ and $y - y'$, i.e.,

$$x - x' = \frac{-mb}{\gcd(a,b)} \qquad \Rightarrow \qquad x' = x + \frac{mb}{\gcd(a,b)}$$
$$y - y' = \frac{ma}{\gcd(a,b)} \qquad \Rightarrow \qquad y' = y - \frac{ma}{\gcd(a,b)}$$

for some integer m. Conversely, every choice of an integer m and a solution (x, y) gives a new solution $(x(m), y(m))$ since

$$ax(m) + by(m) = a\left(x + \frac{mb}{\gcd(a,b)}\right) + b\left(y - \frac{ma}{\gcd(a,b)}\right) = ax + by = c$$

So starting with the solution given by the Euclidean algorithm, which we can call the $m = 0$ solution, we get another solution for each nonzero integer value of m.

Here is a simple, though rather silly, example. I have a certain number of \$5 bills and a certain number of \$2 bills, and I have \$5 in all. What bills do I have? Phrasing this as

$$5f + 2t = 5$$

and handing it to the Euclidean algorithm, I seem to have 5 five-dollar bills and -10 two-dollar bills, i.e., $f = 5$, $t = -10$! Of course we can fix this up by trading in 2 five-dollar bills for 5 two-dollar bills, twice. This is the $m = -2$ solution, $f(-2) = 5 - 2 \cdot 2 = 1$, $t(-2) = -10 + 2 \cdot 5 = 0$. Still one wonders why the Euclidean algorithm gives this strange result.

It turns out the Euclidean algorithm always produces the (x, y) nearest the origin, i.e., the $m = 0$ solution is the "smallest" solution (minimal solution) in the sense of minimizing $x^2 + y^2$. In the above example, the Euclidean algorithm solves

$$5x + 2y = \gcd(5, 2) = 1$$

and the solution it finds, $(x, y) = (1, -2)$, is closer to (0,0) than other ones, like the $m = 1$ solution, $(3, -7)$, or the $m = -1$ solution, $(-1, 3)$. We then scale by the factor 5, $(f, t) = (5x, 5y)$ (according to equation (3.1)), and that is where that first solution comes from. (Note that the scaled solution is *not* the minimal solution to the *original* problem though! In fact the minimal solution is the one we wanted.) Verify in a few other examples that the Euclidean algorithm produces the minimal solution.

Question 7: Can you think of a proof that the Euclidean algorithm produces the minimal solution?

Here is an outline of a proof which you can fill in.

1. Show that the Euclidean algorithm produces the same solution (x, y) to $ax + by = c$ if you divide both sides through by $\gcd(a, b)$, so that it is enough to consider the case $\gcd(a, b) = 1$. You can verify this with computer examples, but of course that is not a proof. On the other hand looking at what the computer is doing may lead you to a proof.

2. Let $(x(m), y(m))$ denote the shifted solutions as above and let (x, y) denote the solution produced by the Euclidean algorithm (which could be thought of as $(x(0), y(0))$). Show that $x^2(m)+y^2(m)$ is a quadratic function of m, and hence that it would be enough to show that (x, y) is no farther from the origin than $(x(1), y(1))$ and $(x(-1), y(-1))$. Show that this in turn is equivalent to

$$|2ay - 2bx| \leq a^2 + b^2 \tag{3.2}$$

3. Strategy: if we could show

$$|2y| \leq a, \qquad |2x| \leq b \tag{3.3}$$

then equation (3.2) would follow by the triangle inequality (show this). You might verify that this strategy is promising by checking whether equation (3.3) is true in examples.

4. Since $b = 1$ is a trivial case, we can assume $1 < b < a$. Show that if the Euclidean algorithm finishes in exactly 1 step, (i.e., if we get remainder $r = 1$ in the first step) then the inequalities in equation (3.3) are true.

5. The magic of induction! On the assumption that the inequalities in equation (3.3) are true for any integer pair that requires N steps in the

Euclidean algorithm, show that they are also true for a pair that requires $N+1$ steps. Do this by noting that the transition from $\gcd(a,b)$ to $\gcd(b,r)$ in the basic step of the Euclidean algorithm reduces the number of steps by 1. Thus, by the induction hypothesis you know

$$|2y'| \leq b, \qquad |2x'| \leq r$$

in the Euclidean algorithm solution to

$$bx' + ry' = 1$$

Now work back one step, using $a = qb + r$.

3.5 Additional topic

Throughout this chapter we have been talking about the arithmetic of the integers, but usually we just wrote symbols like a, b, ..., and just used certain properties, for example that we could multiply, add, that there was an order relation $<$, etc. Does any of this make sense if a and b stand for something else? Look at the case of polynomials in one variable x, since there is a "long division" algorithm for polynomials which, for any polynomial pair (a,b), produces a quotient polynomial q and remainder polynomial r such that $a = qb+r$ holds, with the *degree* of r less than the *degree* of b. Is there a Euclidean algorithm for polynomials? Does it make sense to talk of the gcd of two polynomials, and would this algorithm find it?

3.6 Computer programs

3.6.1 *MATLAB/Octave* programs

Euclid1

```
function [as,bs]=euclid1(a,b)
%function [as,bs]=euclid1(a,b)
%
%Applies Euclidean algorithm to integers a and b
%
%Inputs
%  a - integer to apply Eucliden Algorithm to
%  b - other integer to apply Eucliden Algorithm to
%      Algorithm assumes a>b, if not code will swap a and b
%Output
%  as - vector of sequence of a values
%  bs - vector of sequence of b values

if a<b %if a<b swap a and b
  as=b; %initialize "as" vector with b
  bs=a; %initialize "bs" vector with a
else   %no need to swap a and b
  as=a; %initialize "as" vector with a
  bs=b; %initialize "bs" vector with b
end; %if

i=1; %itialize variable to count step
while bs(i)>0 %repeat until b is 0
  as(i+1)=bs(i);  %next a is previous b
  bs(i+1)=rem(as(i),bs(i)); %next b is remainder of a/b
  i=i+1; %increment step count
end; %while
```

Euclid2

```
function [thegcd,numsteps,x]=euclid2(N)
%function [thegcd,numsteps,x]=euclid2(N)
%      or
%         [thegcd,numsteps]=euclid2(N)
```

```
%
%Applies Euclidean algorithm to two random integers between 1 and N
%
%Input
%  N - a positive integer
%Outputs
% thegcd - the gcd of the two random integers between 1 and N
% numsteps - the number of steps of the Euclidean algorithm required
% x - the two random integers selected (this output is optional)

x=floor(1+rand(2,1)*N); %creates a 2x1 vector of random integers
[as,bs]=euclid1(x(1),x(2)); %applies euclidean algorithm
numsteps=length(as)-1; %number of steps is the number of
                       %entries in "as" minus 1
thegcd=as(end); %the last entry of "as" is the gcd
```

Euclid3

```
function freq=euclid3(N,M)
%function freq=euclid3(N,M)
%
%Applies Euclidean algorithm to M pairs of random integers
%between 1 and N and gives frequencies of number of steps
%required.
%
%Inputs
%  N - maximum value of possible random integers
%  M - number of random pairs to apply Euclidean algorithm to
%Output
%  freq - 1x20 vector representing the frequency of occurence
%         of number of steps The ith entry gives frequency of
%         i steps (i between 1 and 20)

freq=zeros(1,20); %initalize freq as a 1x20 vector of zeros
```

```
for i=1:M  %repeat M times
   [thegcd,numsteps,x]=euclid2(N); %apply Euclid2 code
   if numsteps<=20  %frequency only tracked for
      freq(numsteps)=freq(numsteps)+1; %increment the "numsteps"
                                       %entry of freq by 1
   end; %if
end; %for
```

Euclid4

```
function freq=euclid4(N,M)
%function freq=euclid4(N,M)
%
%Applies Euclidean algorithm to M pairs of random integers
%between 1 and N and gives frequencies of gcd found.
%
%Inputs
%  N - maximum value of possible random integers
%  M - number of random pairs to apply Euclidean algorithm to
%Output
%  freq - 1x20 vector representing the frequency of occurence
%          of number of steps The ith entry gives frequency of
%          gcd=1 (i between 1 and 20)

freq=zeros(1,20);
for i=1:M
   [thegcd,numsteps,x]=euclid2(N);
   if thegcd<=20
      freq(thegcd)=freq(thegcd)+1;
   end; %if
end; %for
```

Euclid5

```
function []=euclid5(a,b)
%function []=euclid5(a,b)
%
%Applies Euclidean algorithm to integers a and b to solve
%the equation
%         a x + b y = gcd(a,b)
%
%Inputs
%  a - integer to apply Eucliden Algorithm to
%  b - other integer to apply Eucliden Algorithm to
%      Algorithm assumes a>b, if not code will swap a and b
%Output
%  none - output is print to screen

if b>a %if b>a, swap a and b
   c=b;
   b=a;
   a=c;
end; %if
a1=a; %save value of "a" input
b1=b; %save value of "b" input

%initalize variables for x,y.

%xold, yold represent x,y values from previous step
xold=1; %x_{-1} = 1
yold=0; %y_{-1} = 0

%x,y represent current x,y values
x=0;  %x_0 = 0
y=1;  %y_0 = 1

%initalize remainder
r=1;
```

```
while r!=0 %loop while remainder nonzero
   q=floor(a/b); %quotient
   r=rem(a,b);   %remainder
   if r==0 %done when remainder is 0
      %output equation
      fprintf('GCD = %d = %d*%d + %d*%d\n',b,a1,x,b1,y)
   end; %if
   xnew=xold-q*x;  %iterate x
   ynew=yold-q*y;  %iterate y
   %shift current values to old values
   xold=x;
   yold=y;
   %shift new values to current values
   x=xnew;
   y=ynew;
   a=b; %update a
   b=r; %update b
end; %while
```

3.6.2 *Mathematica* programs

Euclid1

"Euclid1" computes the GCD of two positive integers. The output shows and counts all intermediate steps.

```
Euclid1[a_, b_]:=Module[{c = a, d = b, ctr = 1}, {c, d} = Sort[{c, d}];
Print["Step ", ctr, ":\t ", c, "\t", d];
While[c ≠ 0, {c, d} = {Mod[d, c], c};
   ctr++;
   Print["Step ", ctr, ": \t", c, "\t", d]]]
```

Here is an example:

Euclid1[24, 15]

Step 1:	15	24
Step 2:	9	15
Step 3:	6	9
Step 4:	3	6
Step 5:	0	3

Here is the first example in the chapter:

Euclid1[6923, 5117]

Step 1:	5117	6923
Step 2:	1806	5117
Step 3:	1505	1806
Step 4:	301	1505
Step 5:	0	301

Euclid2

"Euclid2" picks two random integers and computes their GCD. The rest works the same as "Euclid1". "Euclid2" takes one integer input: N. The two numbers whose GCD is being computed are chosen at random between 1 and N.

```
Euclid2[N_Integer]:=Module[{c = RandomInteger[{1, N}],
d = RandomInteger[{1, N}], ctr = 1}, {c, d} = Sort[{c, d}];
Print["Step ", ctr, ":\t ", c, "\t", d];
While[c ≠ 0, {c, d} = {Mod[d, c], c};
```

```
ctr++;
Print["Step ", ctr, ": \t", c, "\t", d]]]
```

Execute the command below several times to check that different pairs of numbers between 1 and 10^9 =1,000,000,000 are picked each time. Step 1 lists the two random integers chosen.

```
Euclid2[10^9]
```

Step 1:	447488405	616213718
Step 2:	168725313	447488405
Step 3:	110037779	168725313
Step 4:	58687534	110037779
Step 5:	51350245	58687534
Step 6:	7337289	51350245
Step 7:	7326511	7337289
Step 8:	10778	7326511
Step 9:	8249	10778
Step 10:	2529	8249
Step 11:	662	2529
Step 12:	543	662
Step 13:	119	543
Step 14:	67	119
Step 15:	52	67
Step 16:	15	52
Step 17:	7	15
Step 18:	1	7
Step 19:	0	1

Euclid3

"Euclid3" takes two positive integers as inputs, n and M. The program picks M pairs of randomly chosen numbers between 10^n and 10^{n+1} and computes how many steps are necessary in the Euclidean Algorithm to compute their GCD. The output lists how many times a certain number of steps was required to compute the GCD. The information is also listed as a bar graph. **Make sure you understand the meaning of the output!** Also, while M can be rather large, keep n reasonably small.

```
Euclid3[n_Integer, M_Integer]:=(Print["Frequency count for ", M, " pairs of
random integers chosen between ", "10"^n, " and ", "10"^(n+"1")];
store = {};
Do[Module[{c = RandomInteger [{10^n, 10^(n+1)}] , d =
RandomInteger [{10^n, 10^(n+1)}] , ctr = 1}, {c, d} = Sort[{c, d}];
   While[c ≠ 0, {c, d} = Sort[{Mod[d, c], c}];
   ctr++; ];
   z = ctr;
   store = Append[store, z]; ]; , {M}];
Print[TableForm[Sort[Tally[store]],
TableHeadings → {None, {"Number of Steps", "Frequency"}}]];
poss = Range[Max[store]];
Print[BarChart[(Count[store, #]&/@poss), ChartLabels → poss,
AxesLabel → {"Number of steps", "Frequency"}]])
```

Euclid3[3, 50]

Frequency count for 50 pairs of random integers chosen between 1000 and 10000

Number of Steps	Frequency
4	1
6	1
7	9
8	13
9	8
10	10
11	5
12	2
13	1

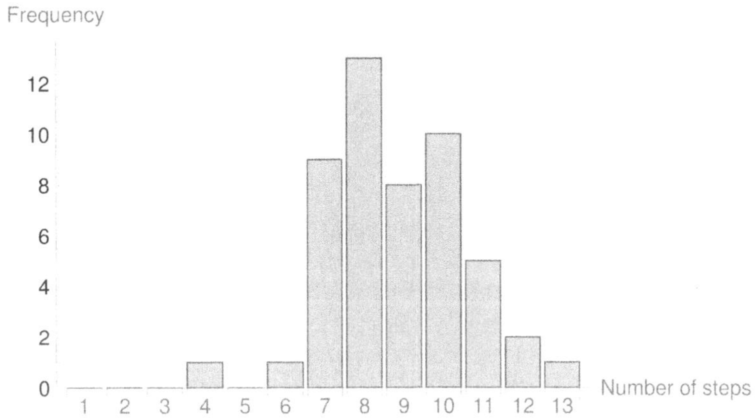

Euclid4

"Euclid4" is similar to "Euclid3", it takes two positive integers as inputs, n and M. The program takes M pairs of randomly chosen numbers between 10^n and 10^{n+1} and computes their GCD. The output lists how many times a certain GCD was encountered. The information (for the 20 smallest GCDs only) is also listed as a bar graph. **Make sure you understand the meaning of the output!** Also, while M can be rather large, keep n reasonably small.

```
Euclid4[n_Integer, M_Integer]:=(Print["GCD distribution for ", M, " pairs of
random integers chosen between ", "10"^n, " and ", "10"^(n+"1")];
store = {};
Do[store = Append[store, GCD[RandomInteger [{10^n, 10^(n+1)}] ,
    RandomInteger [{10^n, 10^(n+1)}]]], {M}];
Print[TableForm[Sort[Tally[store]], TableHeadings → {None, {"GCD", "Frequency"}}]];
Print[BarChart[(Count[store, #]&/@Range[20]), ChartLabels → Range[20],
    AxesLabel → {"GCD", "Frequency"}]])
```

Euclid4[8, 500]

GCD distribution for 500 pairs of random integers chosen between 100000000 and 1000000000

GCD	Frequency
1	311
2	80
3	32
4	17
5	13
6	5
7	5
8	5
10	2
11	1
12	3
13	3
14	2
16	2
17	1
18	1
19	2
20	2
22	1
23	2
25	1
26	1
27	1
52	1
54	1
67	1
83	1
92	1
982	1
11966	1

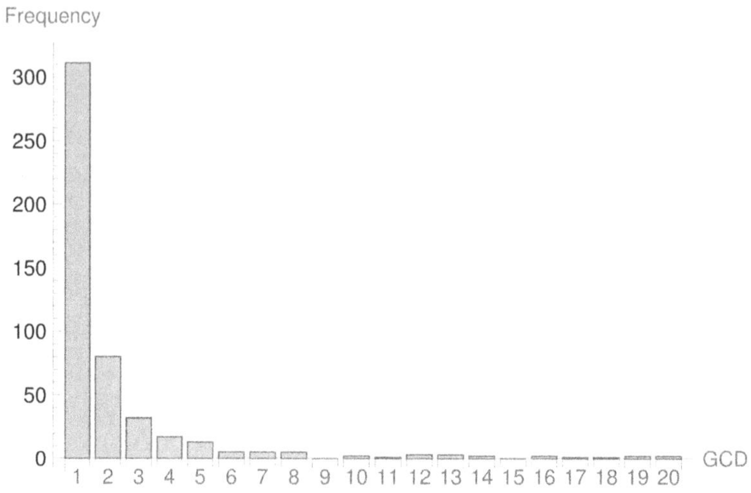

Fibonacci numbers

A table of the first 25 Fibonacci numbers:

$F[1] = 1; F[2] = 1; F[\text{n\_}] := F[n] = F[n-2] + F[n-1]$

TableForm [Table[{n, $F[n]$}, {n, 1, 25}], TableHeadings → {None, {n, "n^{th} Fibonacci number"}}]

n	n^{th} Fibonacci number
1	1
2	1
3	2
4	3
5	5
6	8
7	13
8	21
9	34
10	55
11	89
12	144
13	233
14	377
15	610
16	987
17	1597
18	2584
19	4181
20	6765
21	10946
22	17711
23	28657
24	46368
25	75025

The next command computes the nth Fibonacci number directly:

Fibonacci[6]

8

Fibonacci[100]

354224848179261915075

The following command finds the GCD of the 21st and 24th Fibonacci number:

$k = 21$; Euclid1[Fibonacci[k], Fibonacci[$k + 3$]]

Step 1:	10946	46368
Step 2:	2584	10946
Step 3:	610	2584
Step 4:	144	610
Step 5:	34	144
Step 6:	8	34
Step 7:	2	8
Step 8:	0	2

Linear Diophantine Equations

The command below computes integer solutions to the Linear Diophantine Equation $a\,x + by = \text{GCD}(a, b)$.

Dioph[a_, b_]:=Module[{a2 = Max[a, b], b2 = Min[a, b], a1 = Max[a, b],

b1 = Min[a, b], xold = 1, yold = 0, x = 0, y = 1, c = 1},

```
Print["The GCD of ", a2, " and ", b2, " is ", GCD[a2, b2], ".", "\nWe want to
find integer solutions of ", a2, " x +", b2, " y = ", GCD[a2, b2], "."];
While[c ≠ 0, q = IntegerPart[a2/b2];
c = a2 - qb2;
If[IntegerQ[c], Print[a1x + b1y, " = ", a1, " · (", x, ") + ", b1, " · (", y, ")"], ];
xnew = xold - qx;
ynew = yold - qy;
xold = x;
yold = y;
x = xnew;
y = ynew;
a2 = b2;
b2 = c]]
```

Dioph[15, 24]

The GCD of 24 and 15 is 3.
We want to find integer solutions of $24x + 15y = 3$.

$15 = 24 \cdot (0) + 15 \cdot (1)$

$9 = 24 \cdot (1) + 15 \cdot (-1)$

$6 = 24 \cdot (-1) + 15 \cdot (2)$

$3 = 24 \cdot (2) + 15 \cdot (-3)$

Dioph[6923, 5117]

The GCD of 6923 and 5117 is 301.
We want to find integer solutions of $6923x + 5117y = 301$.

$5117 = 6923 \cdot (0) + 5117 \cdot (1)$

$1806 = 6923 \cdot (1) + 5117 \cdot (-1)$

$1505 = 6923 \cdot (-2) + 5117 \cdot (3)$

$301 = 6923 \cdot (3) + 5117 \cdot (-4)$

Chapter 4

Prime Numbers

4.1 Introduction

The natural numbers, or counting numbers as they are often called, have been objects of human curiosity and investigation since the very beginnings of our intellectual history. Indeed the first rigorous treatment of number theory appears alongside that of geometry in the work of Euclid around 300 B.C.E. In this unit, you will explore patterns of behavior of an especially important group of numbers, the primes. In investigating primes you will

- Look closely at certain of their regularities and irregularities;
- Examine the kinds of questions mathematicians have asked about them; and
- See evidence for some conjectures which can now be proved as theorems and for others whose explanations remain maddeningly elusive.

In the process, you may

- Make some new interesting observations, or
- Discover ways in which you can extend your experimentation to look at variations on the questions posed in this unit.

In the end, perhaps you will find your own curiosity piqued, or even captured, by their intriguing magic.

4.2 Listing prime numbers

Recall that a natural number n is *prime* if it is divisible only by 1 and itself (and if it is not 1). The first few primes are $\{2, 3, 5, 7, 11, 13, 17, ...\}$. You probably remember learning to factor natural numbers in junior high school, and you might even remember asking the question, "How do I know when I'm finished?" The answer, of course, was that you were finished when no single factor could be written as a product of two still smaller ones. In other words, the factoring process ended when you had written that natural number as a product of powers of prime numbers. For example,

$$\begin{aligned}
60984 &= 4 \times 15246 = 2 \times 2 \times 2 \times 7623 \\
&= 2 \times 2 \times 2 \times 3 \times 2541 \\
&= 2 \times 2 \times 2 \times 3 \times 3 \times 847 \\
&= 2 \times 2 \times 2 \times 3 \times 3 \times 7 \times 121 \\
&= 2 \times 2 \times 2 \times 3 \times 3 \times 7 \times 11 \times 11 \\
&= 2^3 \times 3^2 \times 7 \times 11^2.
\end{aligned}$$

Given any integer, we can produce a similar factorization, or decomposition, into products of primes; in this way, the primes are the multiplicative building blocks for our arithmetic.

Now this decomposition has another important property, which you probably remember but hardly notice, namely its uniqueness. No matter how we break 60984 into smaller factors, our final prime decomposition will always be $2^3 \times 3^2 \times 7 \times 11^2$. *Unique factorization*, as it is called, assures us that this is the case for any number—no matter how you factor it, the final expression as a product of primes will always be the same.

Although we often take this property for granted and think it obvious, it is not so easy to prove. In fact, unique factorization is a very powerful property of our own familiar number system, and, as some of you who go on to study number theory will learn, many other arithmetic structures exist which are not blessed with such reliability.

Incidentally, this crucial property of unique factorization would be lost if we allowed 1 to be defined as a prime number since any number of factors of 1 can be written into a product without changing it. So, for example,

$$\begin{aligned} 60984 &= 2^3 \times 3^2 \times 7 \times 11^2 \\ &= 1 \times 2^3 \times 3^2 \times 7 \times 11^2 \\ &= 1^5 \times 2^3 \times 3^2 \times 7 \times 11^2. \end{aligned}$$

To help us in our study of primes, we must first find a method for producing relatively long lists of them. One of the earliest methods works as follows. Begin with a list of natural numbers like the one in the following diagram. First cross out 1. The smallest remaining number is 2, which, by definition, is prime; leave it, but cross out all subsequent multiples of 2 beginning with $2^2 = 4$, thus eliminating all remaining even numbers in the list. Now the smallest remaining number is 3, again a prime. Cross out all subsequent multiples of 3, beginning with $3^2 = 9$ (note that $3 \times 2 = 6$ is already gone because it is also a multiple of 2). The next number left is 5, so delete all subsequent remaining multiples of 5. The smallest number remaining after 5 is 7, so delete all subsequent multiples of 7. Continue in this way, deleting all subsequent multiples of each prime as it appears.

Exercise 1: Try out this method on the following table to create a list of all prime numbers between 1 and 100.

	1	2	3	4	5	6	7	8	9
10	11	12	13	14	15	16	17	18	19
20	21	22	23	24	25	26	27	28	29
30	31	32	33	34	35	36	37	38	39
40	41	42	43	44	45	46	47	48	49
50	51	52	53	54	55	56	57	58	59
60	61	62	63	64	65	66	67	68	69
70	71	72	73	74	75	76	77	78	79
80	81	82	83	84	85	86	87	88	89
90	91	92	93	94	95	96	97	98	99

This ancient method for producing primes has the advantage of utilizing only the simple operation of multiplication. It was developed in the 3rd century BCE by Eratosthenes and is called the Sieve of Eratosthenes, after its creator. Modern versions of this simple sieve are used today to provide powerful estimates of the number of primes with certain properties.

For our purposes, we will find it simpler to use division to generate our lists of primes. Though the operation is inherently more complicated than that of multiplication, the resulting computer program for determining primality is more transparent than one employing sieves. We proceed as follows. Suppose we wish to decide whether or not a given number m in our desired range is a prime. Obviously, we can eliminate immediately all even numbers larger than 2 from our list. Now, if m is odd and not a prime, then $m = ab$ where a and b are integers with $2 < a, b < m$. To avoid duplication of the form $15 = 3 \times 5 = 5 \times 3$, we may assume that $a \leq b$. We can test m for such a non-trivial decomposition by looking at whether the quotient m/a is an integer for some integer a in the range $2 < a < m$. However, we can make this test more efficient by noting that we need to consider only those a such that $a \leq \sqrt{m}$. For if $a > \sqrt{m}$, then $b \geq a > \sqrt{m}$, and $ab > \sqrt{m}\sqrt{m} = m$. Thus, if m/a is an integer for any integer a in the range $2 < a \leq \sqrt{m}$, then m is not prime. As usual, we let $\text{int}(x)$ denote the greatest integer less than or equal to x (so, for example, $\text{int}(3.15) = \text{int}(\pi) = \text{int}(3) = \text{int}(3.999) = 3$ and $\text{int}(-3.15) = \text{int}(-\pi) = -4$). Here is a program that decides whether or not a number is prime.

Program outline: Prime Test

```
Input: an integer n
Output: 1 if n is a prime, 0 otherwise

prime:= 1
a := 2
WHILE prime = 1 AND a < int(sqrt(n)) + 1
      IF n - int(n/a)*a = 0
         prime := 0
      END IF
      a := a + 1
END WHILE
OUTPUT prime
```

Now that we have a program that determines whether a number is prime, here is a program that lists the primes up to a specified integer.

Program outline: List Primes

```
Input: an integer n
Output: a list of primes less than or equal to n

theprimes := [] %an empty dynamic array
i := 1
WHILE m < n + 1
     IF m is prime
        theprimes(i) := m %add m to list of primes
        i := i + 1 %increment i
     END IF
     m := m + 1
END WHILE

OUTPUT theprimes
```

It is perhaps worth mentioning that the first line in the *while* loop in the program above actually calls the program PRIME TEST.

Exercise 2: Examine the corresponding programs at the end of this chapter and explain each step.

Exercise 3: Use this program to produce the set of primes less than 100, and compare it to the result of your hand calculations using the sieve. (Needless to say, the two lists should be identical.)

Exercise 4: Now run the program to produce some large lists of primes. What observations can you make about them? Are there obvious patterns, or do they seem to behave like a fairly irregular collection?

In order to get some feeling for the number of primes, let us define a function which counts the number of prime numbers less than or equal to a number x. This function is traditionally denoted $\pi(x)$ and

$$\pi(x) = \#\{p \leq x : p \text{ prime}\}.$$

For example, $\pi(10) = 4$ since there are exactly 4 primes less than or equal to 10, namely those in the set $\{2, 3, 5, 7\}$. Likewise,

$$\pi(25) = \#\{2, 3, 5, 7, 11, 13, 17, 19, 23\} = 9.$$

The following program is similar to the one above for listing primes, but modified to calculate $\pi(x)$ or, as we will call it in the program, $pi(n)$.

Program outline: Count Primes

```
Input: an integer n
Output: pi(n), the number of primes less than or equal to n

RUN List Primes with input n

%length of theprimes vector gives number of primes
pi :=  length(theprimes)
OUTPUT pi
```

Please be sure that you understand why the program works.

Exercise 5: Use this program to calculate $\pi(100)$, $\pi(1000)$, $\pi(5000)$, $\pi(10000)$, $\pi(50000)$, $\pi(100000)$,and $\pi(1000000)$. If you were to continue these calculations with larger and larger values of n, what do you think would happen to $\pi(n)$? Why?

Exercise 6: Euclid showed that there were infinitely many primes. Have you come across a proof? Can you find one? Here are some questions to get you thinking about a possible proof. Suppose p_1 and p_2 are two prime numbers. Make a new number $q = p_1p_2 + 1$. Could p_1 be a factor of q? Could p_2? Why or why not?

4.3 Functions generating primes

We now have a method for listing and counting primes. However, we might ask whether a formula exists which will produce only primes and all of them.

For example, can one find a non-constant polynomial in one variable with integer coefficients which will generate the whole list of primes? Can we find such a formula if we allow exponentials, more than one variable, and perhaps a larger set of numbers as exponents and coefficients? If a formula like the one we seek does not exist, can we find a simple polynomial or exponential function in one variable which might produce an infinite, if incomplete, list of primes?

Exercise 7: In the 1640's, Mersenne studied numbers of the form $2^p - 1$, p a prime, looking for a formula which generated primes. You can easily duplicate his hand calculations for all primes $p < 10$. What do you discover? Conjecture?

We can extend these calculations using a computer. Here is a program that checks whether $2^p - 1$ is a prime for a given prime p.

Program outline: Mersenne Check

```
Input: a prime number p
Output: 1 if  2^p - 1 is prime, 0 otherwise

PP := 0
IF n = 2^p -1 is prime
  PP := 1
END IF
OUTPUT PP
```

Exercise 8: Use the program to continue Mersenne's calculations for $p < 40$. Because of the limitations of the programming language you are using, you may get an overflow as p approaches 40. Of the primes for which you could obtain an answer, what percentage of the resulting numbers are prime?

In 1772 Euler looked for simple polynomials in one variable with integer coefficients which might generate primes. Of course, a reducible polynomial, that is, one which can be factored into two polynomials of smaller degree with integer coefficients, will never produce primes (why?); hence, only irreducible polynomials could be candidates. Euler's most famous example is the quadratic polynomial $x^2 + x + 41$. The following is a modification of the program above for Mersenne numbers which will let you reproduce Euler's calculations.

Program outline: Euler Check

```
Input: an integer k
Output: integers m from 0 to k for which m^2+m+41 is prime


ms := [0,1,2,...,k]  %vector of m values
ns := m.^2 + m + 41 %vector with n(i)=m
mlist : = vector of ms(i) for ns(i) is prime

OUTPUT mlist
```

Here the symbol .^ indicates element-wise exponentiation of a vector, so for example $m.\hat{}\,2$ would be a vector whose elements are the elements of m squared.

Exercise 9: Run the program for $k = 10, 20, 30, 40, 50$. What observations do you make?

Exercise 10: Modify the program to test the polynomial $x^2 - 79x + 1601$ for for $k = 10, 20$, 30, 40, 50, 60, 70, 80 and 90 ... What do you observe in this case? (Remember to enter the polynomial as $m^2 - 79m + 1601$ in your modified program.) How do your results compare to those for Exercise 8? Can you explain what you observe?

Exercise 11: Now, by hand, evaluate the polynomial $x^2 + x + 2$ for small integer values of x, say $0 \leq x \leq 7$. What is the longest string of primes it yields and for which values of x do they appear? Can you write a proof which explains your observations?

Exercise 12: Modify the program to test various irreducible quadratic polynomials of the form $x^2 + x + q$, for $q = 3$, 5, 11, 17, and 41 and $k \leq 50$. For each q fill out the following table:

q	length of longest string of primes	values of m which yield the primes in the longest string
2		
3		
5		
11		
17		
41		

We have chosen the numbers q above rather carefully. Modify your program to see what happens if $q = 7$ or $q = 37$.

With the small set of experiments above in mind, you might ask yourself whether there exists a formula whose output is exactly the whole set of prime numbers or whether there exists a formula which takes on only prime values, even if it does not produce the complete list of primes. Do other related questions arise as you examine your results?

We have some sense from our results above that the problem of finding simple formulas which take on only prime values is a difficult one. Suppose we pose a somewhat different question: Do simple formulas exist which yield an infinite number of primes among their output? The answer to this one is a resounding yes. In fact, we all know that the arithmetic progression $\{1 + 2k \mid k \text{ an integer}\}$ contains the complete set of odd integers and that among these lie all the prime numbers except 2. On the other hand, the progression $\{0+2k \mid k \text{ an integer}\} = \{2k \mid k \text{ an integer}\}$ contains the single prime number 2. Hence, the polynomial $2x+1$ generates an almost complete list of primes as x takes on all positive integer values, while the polynomial $2x$ generates only one.. In looking more closely at the list of odd numbers, we see that each one lies either in the progression $\{1 + 4k \mid k \text{ an integer}\}$ or in the progression $\{3 + 4k \mid k \text{ an integer}\}$. Let us examine these to see how the primes divide themselves between the two progressions. Do they distribute themselves unevenly, as in the case of the progressions above with difference 2, or do they fall more equally into each category? Let

$$\pi_1(n) = \#\{\text{primes } p \mid p \leq n \text{ and } p = 1 + 4k \text{ some positive integer } k\}$$

and

$$\pi_3(n) = \#\{\text{primes } p \mid p \leq n \text{ and } p = 3 + 4k \text{ some positive integer } k\}$$

Exercise 13: Find an expression relating $\pi(n), \pi_1(n)$, and $\pi_3(n)$.

Exercise 14: Calculate by hand $\pi_1(25)$, $\pi_3(25)$, $\pi_1(50)$, $\pi_3(50)$,$\pi_1(75)$, $\pi_3(75)$, $\pi_1(100)$, and $\pi_3(100)$.

Here is a modification of of the program for finding $\pi(n)$ to evaluate $\pi_1(n)$ and $\pi_3(n)$. The name of the program reflects the language commonly used to describe the arithmetic progressions $\{m+4k \mid k$ an integer $\}$ for $m = 0, 1, 2, 3$. We say an integer is **congruent to m mod 4** if it is in the arithmetic progression $\{m + 4k \mid k$ an integer $\}$. For example, 15 is congruent to 3 mod 4, and 130 is congruent to 2 mod 4. Using this language, $\pi_1(n)$ is the number of primes less than or equal to n that are congruent to 1 mod 4, and $\pi_3(n)$ is the number congruent to 3 mod 4.

Program outline: Count Primes Mod 4

```
Input: an integer n
Output: pi1(n) and pi3(n)

RUN List Primes with input n and output theprimes
theprimesmod4 := mod(theprimes,4)
pi1 := number of elements of theprimesmod4 which equal 1
pi3 := number of elements of theprimesmod4 which equal 3
```

Exercise 15: Use the program above to fill in the following table.

n	$\pi(n)$	$\pi_1(n)$	$\pi_3(n)$	$\pi_1(n)/\pi(n)$	$\pi_3(n)/\pi(n)$
25					
50					
100					
1000					
10000					
100000					

What conjectures might you make on the basis of this evidence about the way the primes divide themselves between these two progressions? Do your observations suggest that an infinite number of primes lie in each? Why?

Exercise 16: If you came up with a proof that there are infinitely many primes, can you adapt it to show that there are infinitely many primes in one of the two arithmetic progressions above? Here are some questions to get you started. If a and b are any two integers, what are $4ab+1$ and $4ab+3$ congruent to mod 4? If a and b are two integers and you know what each of them is congruent to mod 4, can you determine what ab is congruent to mod 4? Going the other way, if you know what ab is congruent to mod 4, can you say anything about what the integers a and b might be congruent to mod 4?

4.4 Distribution of primes

Another set of questions concerns how primes are distributed. Do they become denser or sparser as they proceed out the number line? To answer this, modify your program for calculating $\pi(n)$ as follows so that you can count the number of primes in various intervals of equal length.

Program outline: Count Primes Mod 4

```
Input: two integers  k, n  (2 < k < n)
Output: the number pi(k, n) of primes in the interval [k, n]

RUN Count Primes with input n and output pin
RUN Count Primes with input m-1 and output pikm1

pikn := pin - pikm1

OUTPUT pikn
```

Exercise 17: Now calculate $\pi(k,n)$, the number of primes in the interval $[k,n]$, for the ranges $[1,100]$, $[1001,1100]$, $[10001,10100]$, and finally $[100001,100100]$. Do you get more or fewer primes in these intervals of length 99 as they move out the number line? Try the same experiment with some other intervals of larger length.

Given a prime, can we know how far to go to the next one? More generally, can we say something about the differences, or *gaps*,between consecutive primes?

Exercise 18: Enumerate the list of primes less than 1000 by $p_1 = 2$, $p_2 = 3$, $p_3 = 5$, $p_4 = 7$, etc., and label each consecutive distance as follows:

$$\begin{aligned} d_1 &= p_2 - p_1 = 3 - 2 = 1, \\ d_2 &= p_3 - p_2 = 5 - 3 = 2, \\ d_3 &= p_4 - p_3 = 7 - 5 = 2, \\ &\vdots \\ d_k &= p_{k+1} - p_k. \end{aligned}$$

You can simply list these differences, or you can look at them geometrically by plotting the points (p_k, d_k) on a graph. What differences appear? How many times does each appear? What other observations do you make. Do you have some conjectures about the gaps between consecutive primes as the primes grow larger? Can you write a simple program which would let you examine some of your conjectures for a much larger list of primes?

Exercise 19: If we cannot find a formula which produces all the primes less than a given number, x, can we at least find one which will tell us how many there are? Use your previous calculations of $\pi(n)$ to fill in the following table. (Note: by $\log n$, we mean the logarithm to the base e, which is often written $\log_e n$ or $\ln n$.)

n	$\pi(n)$	$\pi(n)/n$	$\pi(n)/\sqrt{n}$	$\pi(n)/\log n$	$\pi(n)/(n/\log n)$
10					
100					
1000					
10000					
100000					

What do you conjecture the limit of each ratio will be as n increases? What does that mean about $\pi(n)$ compared to n, $n/\log n$, and $\sqrt{n}$ as n increases?

(If you have access to a good graphing package, you may want to try to plot your results above.)

Incidentally, the first estimate for the way in which the number of primes grows was conjectured by Gauss in 1792. However, he was not able to prove his estimate. It was eventually proved in 1896 independently by Hadamard and de la Vallée Poussin using methods from the theory of complex variables and is called the Prime Number Theorem. A much more precise estimate, with detailed error estimates, is equivalent to a famous conjecture that Riemann made in 1859. This conjecture, the so-called *Riemann hypothesis*, has resisted all efforts to prove it and is arguably the most famous open problem in mathematics.

Exercise 20: Using the fact that $\pi(n)/n$ is the proportion of numbers less than n which are prime, does your conjecture say that the primes are getting more or less dense as n grows? Use your conjecture to estimate the number of primes between 10000 and 11000 and check the accuracy of this estimate by comparing with the actual number. Use your conjecture to estimate the number of primes between 100000 and 101000.

4.5 Further reading

If, after scrutinizing your list of primes, you find that you cannot see any simple pattern emerging, do not feel downhearted. Nobody else has been able to do that either; the primes appear to be a rather irregularly behaved set of numbers. However, there is a whole body of interesting theorems which describe some of their properties, as well as an enormous number of intriguing questions which arise from closer studies of their behavior.

Here are a few books that you can profitably consult for more information – the second is very much in the spirit of this unit and the third is a classic.

1. Apostol, Tom,M., *An Introduction to Analytic Number Theory*, Springer-Verlag, New York, 1976.

2. Giblin, Peter, *Primes and Programming: An Introduction to Number Theory with Computing*, Cambridge University Press, Cambridge, 1993.

3. Hardy, G.H. and Wright, E.M., *An Introduction to the Theory of Numbers*, Oxford University Press, Oxford, 1938 (first edition), 1979 (fifth edition).

4. Ribenboim, Paulo, *A Book of Prime Number Records*, Springer-Verlag, New York, 1988 (revised 1990).

The article by Paulo Ribenboim also deals further with many of the issues that your investigations will have raised: Ribenboim, Paulo, "Prime Number Records," *The College Mathematics Journal*, **25** (1994), 280-290.

4.6 Computer programs

4.6.1 *MATLAB/Octave* programs

Prime Test

```
function p=primetest(n)
%function p=primetest(n)
%
%Determines if a given number is prime
%
%Inputs
%  n - a positive integer
%Output
%  p - indicates if n is prime
%      p=1 if n is prime p=0 n is not prime

if n<2
   p=0; %1 is is not prime
else
   p=1; %assume n prime until we find a divisor
end; %if
```

```
a=2; %a possible divisor of n
while p==1 && a<=sqrt(n)
   if n-floor(n/a)*a == 0
      p=0; %n is not prime
   end; %if
   a=a+1; %increment a
end; %while
```

We have included this program for interest, but *MATLAB/Octave* includes a function ISPRIME which will accomplish the same task, and can also be applied to a vector of numbers.

List Primes

The *MATLAB/Octave* function PRIMES performs the same procedure as described in the pseudocode. To get a list of primes lesser or equal to n and save it as a vector *myprimes* enter

```
>> myprimes=primes(n)
```

Count Primes

indexCount Primes@COUNT PRIMES

```
function primecount=countprimes(n)
%function primecount=countprimes(n)
%
%Gives count of primes less or equal to given number n
%
%Inputs
%  n - a positive integer
%Output
%  primecount - count of primes less or equal to n

primecount=length(primes(n));
```

Mersenne Check

```
function PP=mersennechk(p)
%function PP=mersennechk(p)
%
%Determines if 2^p-1 is prime for a given a positive integer p
%
%Inputs
%  p - a positive integer less than 65
%Output
%  PP - indicates if 2^p-1 is prime
%       PP=1 if 2^p-1 is prime, PP=0 if 2^p-1 is not prime

if p>64
error('Input p must be less than 65')
end; %if

PP=isprime(2^p-1);
```

Euler Check

```
function mlist=eulerchk(k)
%function mlist=eulerchk(p)
%
%Gives list of integers m for which m^2+m+41 is prime
%for m up to a given number k
%
%Inputs
%  k - a positive integer
%Output
```

```
% mlist - list of integers m for which m^2+m+41 is prime

ms=0:k;
mlist=ms(isprime(ms.^2+2*ms+41));
```

Count Primes Mod 4

```
function [pi1,pi3]=primesmod4(n)
%function [pi1,pi3]=primesmod4(n)
%
%Gives counts of primes that are less or equal to given number n
%that are congruent to 1 mod 4, and 3 mod 4.
%
%Inputs
% n - a positive integer
%Output
% pi1 - count of primes less or equal to n that are congruent to
%       to 1 mod 4
% pi3 - count of primes less or equal to n that are congruent to
%       to 3 mod 4

theprimesmod4=mod(primes(n),4);
pi1=sum(theprimesmod4==1);
pi3=sum(theprimesmod4==3);
```

Count Primes in Interval

```
function primecount=countprimesint(k,m)
%function primecount=countprimesint(k,m)
%
%Gives count of primes in interval [k,m]
```

```
%
%Inputs
%  k,m - positive integers
%Output
%  primecount - count of primes betweeen k and m, inclusive

primecount=countrimes(m)-countprimes(k-1);
```

4.6.2 *Mathematica* programs

PrimeTest

PrimeQ[k] answers the question: Is k a prime number?

PrimeQ[5]

True

PrimeQ[6]

False

ListPrimes

The command below generates a list of all primes less than or equal to n.

ListPrimes[n_]:=Module[{$k = 1$, list = {}}, While[Prime[k] $\leq n$,

list = Append[list, Prime[k]]; k++]; list]

ListPrimes[10]

$\{2, 3, 5, 7\}$

Count Primes

The next command gives the number of prime numbers less than or equal to n.

PrimePi[5]

3

MersenneCheck

This command checks whether $2^n - 1$ is prime:

MersenneCheck[n_]:=Print[2^$n-1$, " is prime: ", PrimeQ[2^$n-1$]]

MersenneCheck[6]

63 is prime: False

MersenneCheck[607]

531137992816767098689588206552468627329593117727031923199444138200403559860852242739162502265229285668889329486246501015346579337652707239409519978766587351943831270835393219031728127 is prime: True

While many mathematicians believe there are infinitely many Mersenne primes, there are only 51 known Mersenne primes (as of March 22, 2024). See the Great Internet Mersenne Prime Search (https://www.mersenne.org/)

EulerCheck

This command checks for which m the number $m^2 + m + 41$ is prime (for $m = 0, \dots, k$). You can change the polynomial formula #^2+#+41. Just think of # as the variable of the polynomial. (Sorry, this is geeky.)

```
EulerCheck[k_]:=Select[Table[m, {m, 0, k}], (PrimeQ[#^2 + # + 41]&)]
```

```
EulerCheck[20]
```

$\{0, 1, 2, 3, 4, 5, 6, 7, 8, 9, 10, 11, 12, 13, 14, 15, 16, 17, 18, 19, 20\}$

Count Primes modulo 4

There are no primes divisible by 4 and only one that leaves remainder 2 when divided by 4. The two commands below count how many primes less than or equal to n leave remainder 1, and 3, respectively, upon division by 4.

```
Pi1[n_]:=Module[{k = 1, list = {}}, While[Prime[k] ≤ n,
    list = Append[list, Prime[k]]; k++];
    Select[list, (Mod[#, 4] == 1&)]//Length]
```

```
Pi3[n_]:=Module[{k = 1, list = {}}, While[Prime[k] ≤ n, list = Append[list, Prime[k]];
    k++];
    Select[list, (Mod[#, 4] == 3&)]//Length]
```

```
Pi1[111]
```

13

```
Pi3[111]
```

15

Count Primes in an Interval

The last command computes the number of primes in the interval $[k, n]$:

```
PrimePiInterval[k_, n_]:=PrimePi[n] - PrimePi[k] + If[PrimeQ[k], 1, 0]
```

```
PrimePiInterval[5, 13]
```

4

Chapter 5

The Coloring of Graphs

5.1 Introduction

In this laboratory, we develop the concept of *properly coloring* the vertices of a finite graph. This notion has a long history arising first from the famous "Four Color Theorem", which deals with the coloring of a map on a planar surface. We shall introduce the idea with a very different type of problem. In this laboratory, we shall:

- Describe a problem of scheduling of courses in a limited number of time periods.
- Translate this problem to one of coloring the vertices of a finite graph.
- Introduce the *chromatic polynomial* of a finite graph—giving a very general solution to the scheduling problem.
- Find an algorithm for calculating the polynomial of any finite graph.
- Use a computer program to calculate chromatic polynomials for numerous graphs, conjecturing (and in some cases proving) the relationships between the coefficients of the polynomial for a graph and certain geometric properties of the graph.

5.1.1 A scheduling problem

Example: A number of people have signed up for more than one of the six courses to be given in the mathematical sciences at the local Evening College. It is determined that these students *must* take these courses for their programs. You, as the Registrar, are attempting to find a schedule so that each person can take the courses they want. The available evening time slots are 8 o'clock, 9 o'clock, and 10 o'clock. There are plenty of classrooms available.

The courses are: Calculus (C), Statistics (S), Data Structures (D), Numerical Analysis (N), Graphics (G), and Operating Systems (O).

Suppose that the enrollments are as follows (and that all the students have different last names and that each of them could schedule classes at any of the three times):

Calc.	Stat.	Data St.	Num.An.	Graphics	Op.Sys.
Axel	Carlton	Blim	Ames	Bennett	Ames
Eagle	Carter	Conley	Barnes	Carlton	Barnes
Harns	Eckert	Jones	Franck	Conley	Blim
Janes	Harns	Martinez	Frick	Forrest	Forrest
Plum	Jones	Smith	LaMire	Kennan	Janes
Snapper	Smith	Swee	Wu	Lyon	Myers
Stram	Wills	Wu		Weaver	Talbot
Wong				West	
				Wills	

Is it possible for you (the Registrar) to set schedules for the courses in the times available (8, 9, and 10 o'clock) so that all of these students can be accommodated? If so, how do you do it? Could you do this with two meeting times per week, say 8 and 9 o'clock?

See if you can find a satisfactory pattern of meeting times for these courses within the three time slots. Enter course schedules in a chart like the one below.

Times	Courses for this time
8 o'clock	
9 o'clock	
10 o'clock	

How many different satisfactory patterns are there?

The questions above represent two of the most common problems found in the area of combinatorial mathematics: first, there is the question of determining the existence of a solution and, second, that of counting the number of solutions, if any do exist.

5.2 Introduction to the mathematical ideas

5.2.1 Translation to an equivalent problem

Here we look again at the problem just posed with the idea of developing a method (*an algorithm*) for determining the number of possible meeting times for the class. To get at it, we employ an often fruitful technique in mathematics—we draw a picture. In this way, we give a very different flavor to the class scheduling environment. Let's think of a diagram for the classes in which the courses themselves are drawn as dots. We will draw a line between two "courses" if and only if the courses cannot meet at the same hour due to enrollment conflicts. For example, since courses C and S have enrollments in common (Harns is in both classes), we have a line between points C and S of the diagram (Figure 5.1).

Exercise 1: Here we complete the diagram in Figure 5.1. Label 6 points C, S, D, N, G, O and draw a line segment between two of the points if and only if the two corresponding classes have at least one student in common.

In doing Exercise 1, you have drawn a diagram of the situation which is known in mathematics as a *graph*. In graph theory, the dots (courses) are

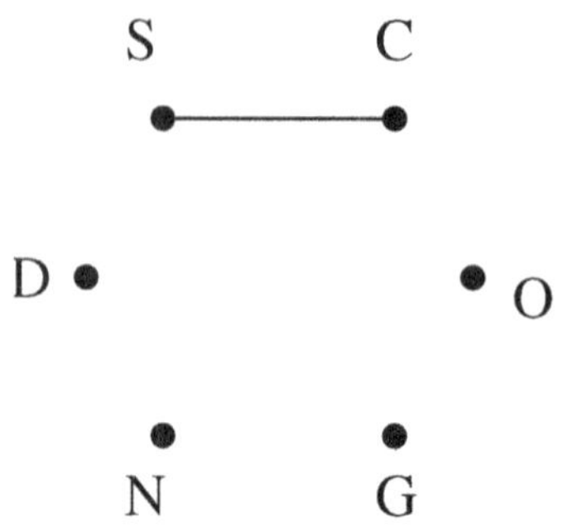

Fig. 5.1: Incomplete graph for course scheduling

called *vertices* or *nodes* of the graph, while the lines which you have drawn are called *lines* or *edges* of the graph. We will deal only with *finite* graphs; that is, graphs having only a finite number of vertices. Now we try a very strange-seeming technique. We will say that we are choosing a *proper coloring* of the graph using n colors (or just a *proper n-coloring*) if we can assign colors to the vertices (or "paint" the vertices) from a "palette" of n different colors in such a way that no two vertices which are joined by an edge are assigned the same color. In this scheme, for instance, vertices C and S must be assigned different colors.

It is appropriate to consider that the different time slots in the scheduling problem can be associated with the different colors. How so? Well, we certainly don't want any pair of courses having students in common (vertices joined by an edge) to be assigned the same time slot (the same color). Also, we want each of the courses to have a time slot (each vertex should get a color). The problem of scheduling courses is, in fact, the same as a graph coloring problem. A proper coloring of the graph corresponds to a satisfactory pattern of meeting times for these six homeless courses.

Thus, we can answer the existence question for our course scheduling by determining whether there is a proper 3-coloring of the graph you drew in Exercise 1. Also, the scheduling counting problem is answered if we can count the number of proper 3-colorings of the graph.

Exercise 2: First try to find a proper coloring of the graph using a palette of two colors. Call the colors A and B. Now suppose you have a palette of three colors, $\{A, B, C\}$. Can you find a proper coloring? (This is the *Existence Problem.*)

Exercise 3: How many proper colorings (if any) can you find using two colors? How many can you find with three colors? (This is the *Counting Problem.*)

5.2.2 The general counting problem

In this section and the next, we will develop an algorithm for finding the number of proper colorings of a graph using n colors. We will find a method that is general enough to work, at least theoretically, with any finite graph. We start with a very simple picture. Take the graph G_1, shown below, in which there are two nodes 1 and 2, and where 1 is joined to 2 with an edge.

The Graph G_1

Suppose there is just one color {red }. Clearly, there is no way to color G_1 with just one color, since 1 and 2 must have different colors. (If there were no edge, we could do it with just one color. Simply color both vertices red and be done with it!)

If we have two colors {red, blue}, then there are two ways to do it:

Exercise 4: Write down all the ways to color the graph G_1 with three colors {red, white, blue}. Write down the ways with four colors $\{A, B, C, D\}$.

Exercise 5: In how many ways (don't write them down!) can you color G_1 with x colors?

Now consider the graph G_2 shown in Figure 5.2.

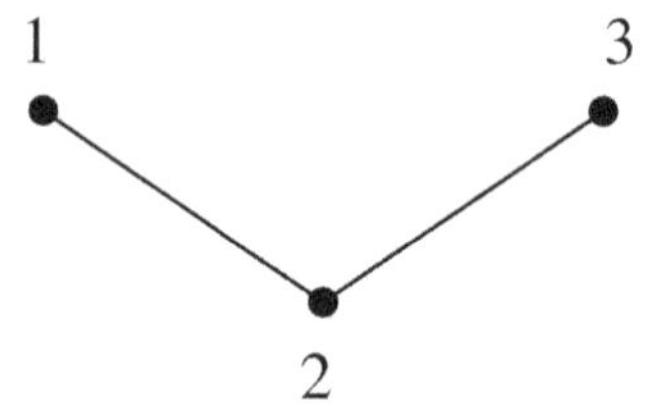

Fig. 5.2: The Graph G_2

Exercise 6: Write down all the ways to color graph G_2 with two colors {red, white}. Write down all the ways with three colors {red, white, blue}.

Exercise 7: A finite graph in which no pairs of vertices are connected by an edge is called an *empty* graph. Find a formula for the number of ways to properly color an empty graph containing n vertices using x colors.

5.2.3 An algorithm for counting colorings

Consider a graph, which we will call G. For a given non-negative integer x, we wish to count the number $P(G;x)$ of possible proper colorings for G using a palette of x colors. As we shall see below, this number can be expressed as a polynomial in x. For this reason, we make the following definition.

Definition: $P(G;x)$, the number of proper colorings of a graph G using x colors, is the *chromatic polynomial* of G.

Look again, for example, at graph G_2. If we have x colors available, we can paint vertex 1 with any one of these colors. This done, we can now paint vertex 2 using any of $x-1$ colors. Hence these vertices together may be colored in $x(x-1)$ ways. Now vertex 3, being adjacent only to vertex 2, may be colored also in $x-1$ ways. Hence, $P(G;x) = x(x-1)^2$. It is tempting to think that in this way we can step through any finite graph and determine its chromatic polynomial!

Exercise 8: Try the technique just described also for graphs G_3, G_4, and G_5 of Figure 5.3. You will note that using this process on G_3 and G_4 is fairly straightforward [yielding, for instance, $x(x-1)(x-2)$ for G_3]. On the other hand, G_5 produces a significant problem. What is it? The lesson here points to the need for a more general approach to calculating chromatic polynomials.

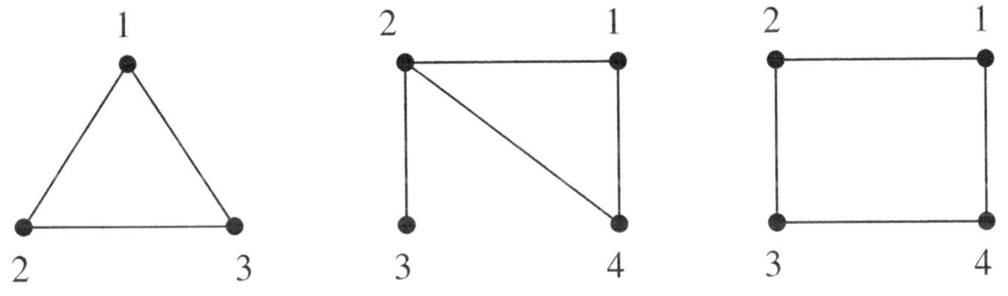

Fig. 5.3: Graphs G_3, G_4, G_5

In thinking about an algorithm for counting colorings, we might focus on what happens if we make a minor change in the graph. Let us take for example the graph H in Figure 5.4 and determine how we would find $P(H;x)$. Notice that H is the same as the graph G_4 in Figure 5.3.

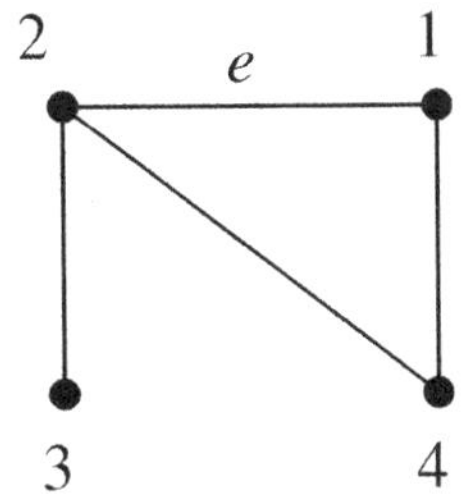

Fig. 5.4: Graph H

To motivate the method, think of another counting problem. Suppose an audience of 200 people have gathered for a concert (you know that number because you've collected that many tickets), and you want to determine the number of those present who did *not* buy refreshments at intermission. It might be easiest to simply count those who *did* buy refreshments and subtract that number from 200. Here we will do the same thing, except instead of counting people, we are counting colorings. To find the number of proper colorings, we will determine a number of colorings including both proper and improper ones and then subtract off the number of improper colorings.

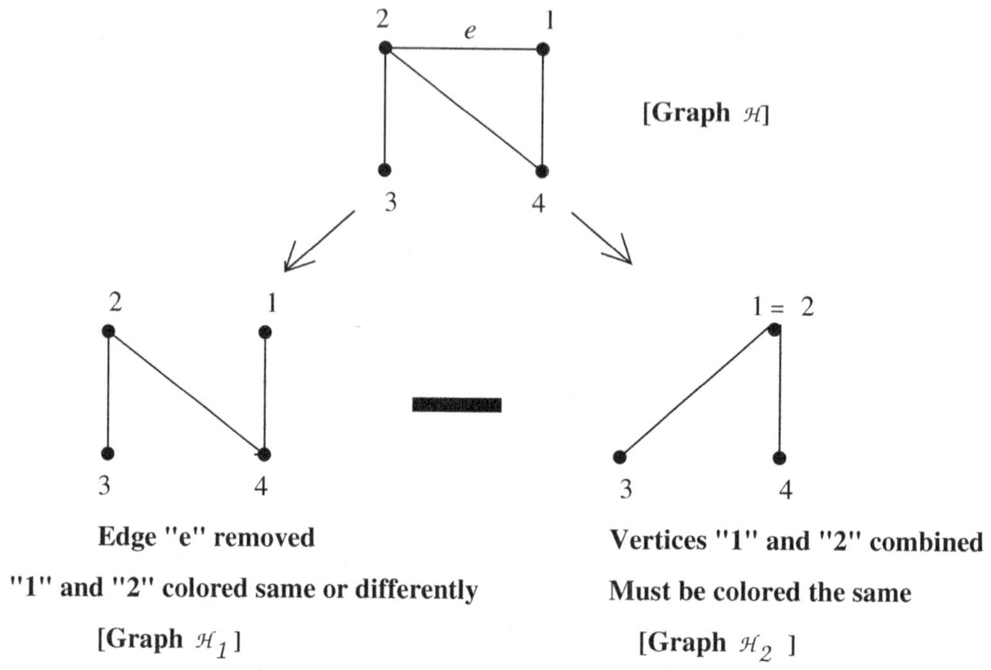

Fig. 5.5: One step in the algorithm

To see how the method works, look at Figure 5.5. We focus on the edge of H labelled e. Each of the ways of *properly* coloring H must have the vertices 1 and 2 colored *differently*. After all, that's what "proper" means.

- If edge e were *not* there (as in graph H_1 of Figure 5.5, in which e has been deleted from H), then vertices 1 and 2 could be colored *either the same or differently*.

- On the other hand, if we were to combine vertices 1 and 2 into a single vertex (as in graph H_2 in Figure 5.5), then we would be assured that they are colored with the *same* color!

Note that in forming the graph H_2 from H by combining the vertices 1 and 2, we put an edge between the new vertex $(1 = 2)$ and another vertex

v in H_2 if and only if there had previously been an edge in H going either between 1 and v or between 2 and v.

Now to the counting. Let p and q be any two vertices of a graph H that are connected by an edge. The number of ways to properly color graph H will be the number of ways of doing so in which p and q are colored with either the same or different colors (i.e., including both improper and proper colorings) *minus* the number of ways of doing so in which they are necessarily colored with the same color (i.e., colored improperly).

This difference is the number of proper colorings of H_1 minus the number of proper colorings of H_2. Symbolically, we can write this:

$$P(H; x) = P(H_1; x) - P(H_2; x) \tag{5.1}$$

Now, however, H_1 and H_2 can each be split up in the same manner, dividing them into graphs $H_{1,1}$, $H_{1,2}$, and $H_{2,1}$, $H_{2,2}$, with a similar formula for each of their numbers of colorings, $P(H_1; x)$ and $P(H_2; x)$. In other words, $H_{1,1}$ is obtained from H_1 by deleting an edge, and $H_{1,2}$ is obtained from H_1 by identifying the two vertices connected by the deleted edge. Similarly for $H_{2,1}$ and $H_{2,2}$. This process continues until the new graphs that are formed have no edges at all (i.e., they're "empty" graphs) and so we cannot continue removing edges. We carry this process out to completion with the graph in Figure 5.6. using G_2 of exercise 6

We know from exercise 7 that an empty graph, say with k vertices, can be colored in x^k ways. Hence, we can gather up all the empty graphs that are formed and count the number of x^k's for each of the corresponding integers k, subtracting off those that ought to be subtracted according to formula (5.1). This procedure, published in 1946, is due to Birkhoff and Lewis. It is called the *Birkhoff-Lewis Reduction Algorithm.*

After the algorithm is carried through, notice that the bottom row of graphs consists of a line of only empty graphs (no edges). From this and from your result of exercise 7, you can conclude that the chromatic polynomial of G_2 is

$$P(G_2; x) = (x^3 - x^2) - (x^2 - x)$$

or, in usual polynomial form:

$$P(G_2; x) = x^3 - 2x^2 + x,$$

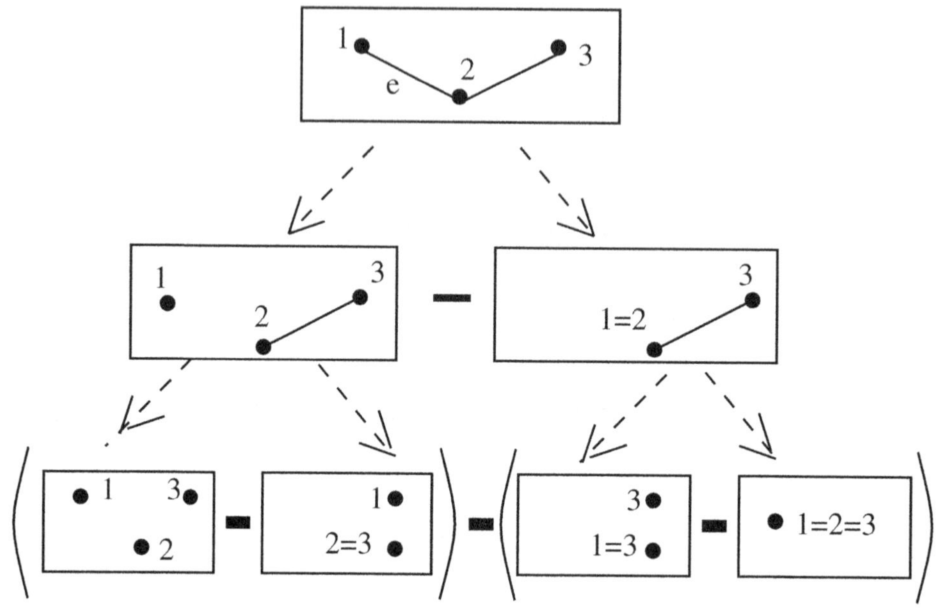

Fig. 5.6: Decomposition of G_2

confirming our earlier calculation.

Notice that all graphs having a finite number of vertices eventually decompose into empty graphs under this process. Hence, the number of ways, $P(G;x)$, to properly color any graph using x colors is going to be a polynomial expression in x having integral coefficients. This follows since, ultimately, $P(G;x)$ consists of sums and differences of expressions of the form x^k.

Exercise 9: Use the procedure outlined above to calculate the chromatic polynomial of the graph H of Figures 5.4 and 5.5. [Answer: $x^4 - 4x^3 + 5x^2 - 2x$—again, confirming earlier calculation.]

Exercise 10: Use the *Birkhoff-Lewis Reduction Algorithm* to find the chromatic polynomial $P(G;x)$ for each of the graphs in Figure 5.7. (The last one is messy; write small!)

The web app CHROMATIC POLYNOMIAL CALCULATOR designed by Thomas Perrot performs the algorithm that you used in Exercise 10. This

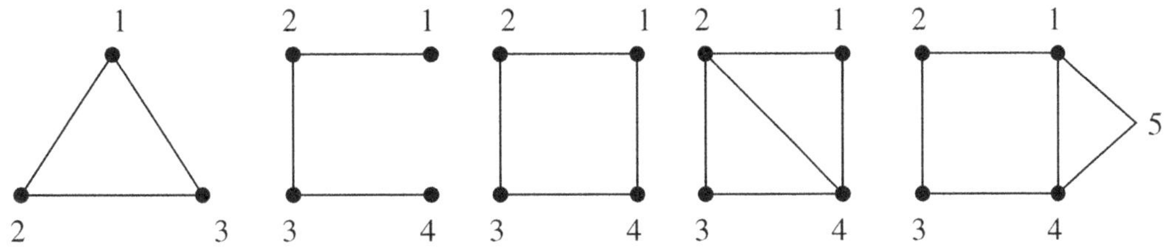

Fig. 5.7: Graphs for Exercise 10

app can be found at: https://tjperr.github.io/chromatic-polynomial/ . Your method involved doing the same thing over and over to graphs derived from the original one. This type of process is called "recursion". A graph is decomposed into two graphs by a certain method, and then the same procedure is applied to those graphs to yield more graphs, repeating until the result consists entirely of empty graphs.

The app has a simple graphical interface, you use the mouse to add vertices and connections between them. When you click the *Calculate* button the computer carries out the recursive calculations until it reduces all of its graphs down to empty graphs, counting these empty graphs as it finds them. Ultimately, the Birkhoff-Lewis reduction algorithm leads us to a count of the number of x^k's in the polynomial, generating the coefficients of the chromatic polynomial. The chromatic polynomial $P(G; x)$ for you graph will be displayed.

Note: The constant term of the chromatic polynomial is always 0. Why does that make sense? What is the value of $P(G; x)$ when $x = 0$? In the chromatic polynomial, the coefficient of x^k is 0 whenever $k > n$. Why is this so?

The value of the polynomial for any positive integer x gives the number of ways to properly color G with x colors. The *least* positive integer value of x which yields $P(G; x) > 0$ is the *smallest* number of colors needed to color the graph. This is called the *chromatic number* of G.

5.3 Questions to explore

Question 1: A graph with n vertices in which all distinct pairs of vertices are connected with edges is called a *complete graph* on n vertices. What are the chromatic numbers for the complete graphs on 2, 3, 4, 5, 6 vertices? [Don't use the computer here—think this one through and clearly describe your reasoning.]

Question 2: What are the chromatic polynomials for the complete graphs on $n = 3, 4, 5, 6$ vertices? [Again, think through how you would do this without the computer. Then use the CHROMATIC POLYNOMIAL CALCULATOR to confirm your conjectures.

Question 3: Using your results from Questions 1 and 2, can you find a general formula for the chromatic polynomial for a complete graph on n vertices?

Question 4: In how many ways can the graph of Figure 5.8 be colored with a pallette of 44 colors? Use the app CHROMATIC POLYNOMIAL CALCULATOR to answer this question. Also, what are the chromatic number and the chromatic polynomial for the graph?

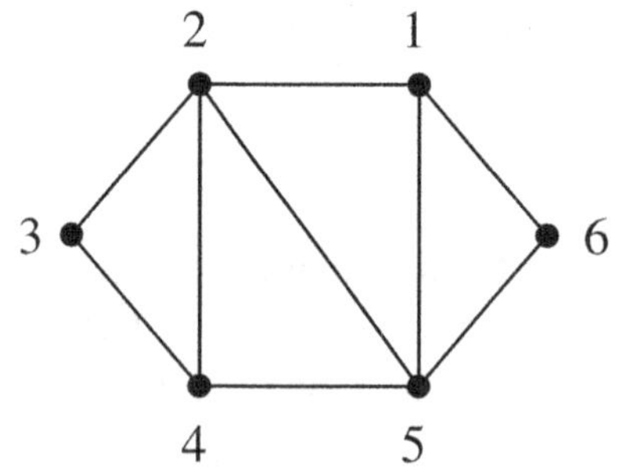

Fig. 5.8: Graph for Question 4

Question 5: Use the CHROMATIC POLYNOMIAL CALCULATOR to find the chromatic polynomials for each of the graphs in Figure 5.9. What are their chromatic numbers? For each graph, can you find a coloring that will use just the chromatic number of colors?

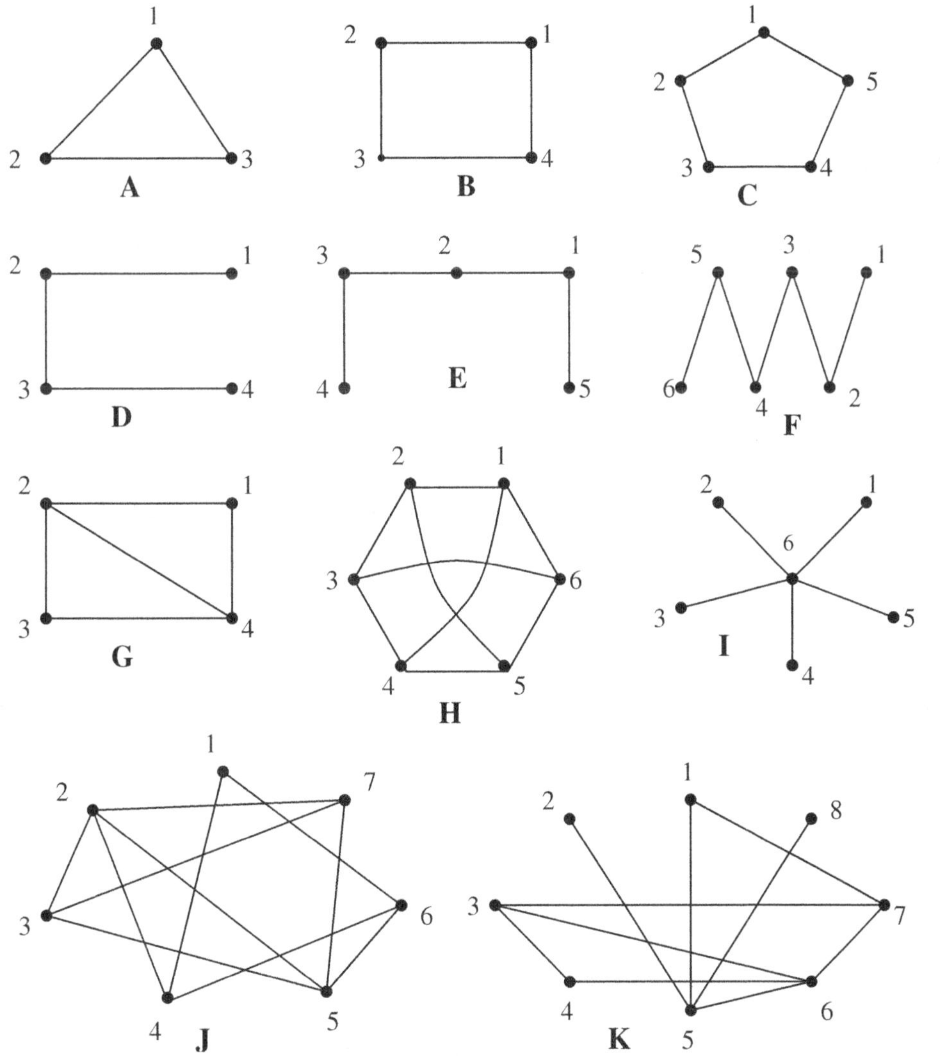

Fig. 5.9: Graphs for Question 5

The first row of graphs used in Figure 5.9 are examples of what are called "cycles" in graph theory. An *n-cycle* consists of a sequence of n distinct vertices v_1, $v_2, \cdots, v_n$ where the edges, and the only edges among them, are v_1 joined to v_2, v_2 to v_3, and so on until, v_{n-1} is joined to v_n, and v_n is joined to v_1. We will define an *n-path* exactly as we did an n-cycle, except that v_n is *not* joined to v_1. The second row of graphs in Figure 5.9 are examples of paths on 4, 5, and 6 vertices, respectively.

Question 6: What is the chromatic number of an n-path? [Reason this one out—and only then look at your results from the computer program.]

Question 7: What is the chromatic polynomial for an n-path? [Again, reason and experiment. Your answers from reasoning and from the computer program may have very different looking forms—but you should be able to see that they are, in fact, the same.]

Question 8: Repeat Question 6 using n-cycles rather than n-paths. In going through the reasoning here, do you run into any snags along the way? What are they?

Question 9: Repeat Question 7 using n-cycles rather than n-paths. Again, what are the snags on this question? You might look again at the Birkhoff-Lewis algorithm for this answer. Be sure to compare your answer with the computer's output.

Question 10: Here's a question to reason through. If the chromatic number of a graph G is k (for some positive integer k), what factors do you know *must* exist for the chromatic polynomial of G?

Question 11: If you have access to a computer algebra program, try factoring the polynomials you have obtained using the CHROMATIC POLYNOMIAL CALCULATOR. What patterns do you observe?

Question 12: We have seen already that the constant term for any chromatic polynomial is always 0 and that the degree of the polynomial is always equal to the number of vertices of the graph G. This question deals with making conjectures as to some of the other ways in which the chromatic polynomial reflects properties of the graph itself. Look carefully at the results of Question 5, and also use the CHROMATIC POLYNOMIAL CALCULATOR to calculate chromatic polynomials of some additional graphs of your own choosing. Can you describe any special characteristics or patterns that you observe in the coefficients of $P(G; x)$. Do you see any pictorial characteristics of the graphs that are reflected in these coefficients? This is a very interesting open-ended question that may involve looking at a large number of graphs with the computer. Have fun with it! After you have spent time exploring these patterns, you can read the article "A Path Less Taken to the Peak of the Math World" about June Huh who recently proved a long standing conjecture about patterns in the coefficients of chromatic polynomials.

References:

Birkhoff, G.D., and Lewis, D.C., Chromatic Polynomials, *Trans. Am. Math. Soc.*, **60** (1946), pp. 355-451.

Roberts, Fred S., *Applied Combinatorics*, Prentice Hall, Englewood Cliffs, N.J., 1984.

Hartnett, K. "A Path Less Taken to the Peak of the Math World". *Quanta Magazine* 2017.
https://www.quantamagazine.org/a-path-less-taken-to-the-peak-of-the-math-world-20170627/

Chapter 6

Randomized Response Surveys

6.1 Introduction

We all use data to estimate unknown probabilities, although often the process is both informal and not something we pay attention to. This chapter introduces elements of a mathematical theory defining properties that distinguish a good estimate from a bad one. Although parts of this theory are more than 300 years old, it remains an active area of research for mathematical statisticians. In what follows, you will:

- Learn a method of estimation called randomized response, invented (less than 60 years ago!) to get reliable answers to sensitive questions;
- Use this method as a context for learning ways to evaluate estimation methods;
- Investigate the relationship between how much data you have and how good your estimate is; and
- Explore ways to improve on the standard method of estimation for randomized response surveys.

6.2 Asking sensitive questions

Have you ever used illegal drugs? If someone were to ask you that, you would be quite justified in regarding the question as intrusive. If a pollster conducting a survey were to ask 1500 randomly selected college students, chances are that some would refuse to answer, and that others would say no, even if in fact they had used drugs. For these reasons, direct questions are unlikely to get reliable information about drug use, or any of a number of other sensitive questions. At the same time, good data are often needed for informed discussion of public policies related to such sensitive issues as abortion, sexually transmitted diseases, drug use, cheating on income taxes, and so on.

One clever way to gather usable data while protecting the privacy of those surveyed was invented by Stanley Warner and published in 1965. Warner's idea was to let each person decide using a chance device whether to answer the question of interest or an innocuous decoy question. Here is a version of Warner's *randomized response* technique:

Imagine that a researcher wants to survey your class in order to find out what fraction of you have ever used illegal drugs. "Have you ever used illegal drugs?" will be his "real" question, and he'll prepare another, "decoy" question to go along with it: he'll ask each of you to toss a penny, secretly, and remember whether it lands Heads or Tails. "Did your penny land Heads?" will be his decoy question. Now for the survey: he'll ask each person to toss a dime, also in secret, and use the result to choose which question to answer.

Results of secret dime toss	Question to answer
HEADS	REAL: Have you ever used illegal drugs?
TAILS	DECOY: Did your penny land Heads?

As long as you don't let the researcher see the results of your coin tosses, and answer only "Yes" or "No," he won't know which question you are answering, and so he won't know whether "Yes" means that you have used drugs, or that your penny landed Heads.

Class survey. This would be a good time to gather class data. First, the class should choose a Yes/No question you'd like to know about. Also, you need to make sure that everyone understands the rules for deciding whether to answer "Yes" or "No." Finally, everyone needs to feel that the method does in fact protect his or her privacy. Once you have carried out the survey and counted the total number of Yes answers, you can work at trying to figure out what that number tells you. In order to study the relationship between the number of Yes answers and the true make-up of the class, it will be handy to have a little background.

6.3 Background

Intuitive definition. Suppose that a well-defined, repeatable chance process generates a numerical quantity Y. The **expected value** of the quantity, written $\mathrm{EV}(Y)$, is the limit of its average value, averaged over repetitions of the chance process, as the number of repetitions increases without bound.

> **Example:** In our case, the chance process corresponds to the randomized response survey applied to your class using your chosen question. This is a process that you can imagine repeating a very large number of times, each time resulting in a different value of $Y =$ total number of Yes answers, depending on how the coins land. If you repeat the survey 100 times, and compute the average number of Yes answers, that average will be a good estimate for the expected value of the number of Yes answers, or $\mathrm{EV}(Y)$. If you repeat the survey 1000 times, the average will be a better estimate; 10,000 repetitions would give an even better estimate. In principle, you can imagine taking the limiting value of the average as the number of repetitions goes to infinity; that number is the expected value.
>
> **History:** In classical probability theory, the chance process and its expected value are defined abstractly, without reference to any actual physical mechanism, and there is a theorem (called the Weak Law of Large Numbers, and first proved by Jacob Bernoulli

in 1692) that the limiting long-run average equals the expected value.

General principle for statistical thinking. To interpret the outcome of a repeatable process, assume that the value you observe is equal (at least approximately) to its expected value.

Rationale. For a great many chance mechanisms—but not all!—outcomes near the expected value are more likely than outcomes far from the expected value. The principle is most reliable when the numerical outcome in question is the sum of a very large number of very small chance-like components. (The basis for this last statement is another theorem of classical probability theory, called the Central Limit Theorem, due mainly to the work of De Moivre in the 1720s and Laplace in the 1750s.)

Application. The principle will lead to a way to go from the observed number of Yes responses to an estimate for the fraction of true Yes people in the class: assume the observed number of Yes answers equals the expected number of Yes answers, and solve the resulting equation. The expected number of Yes answers will be an expression involving the number of true Yes people, and so solving will give you an expression for the number of true Yes people in terms of the observed number of Yes answers. To make this strategy work, you need to find an expression for the expected number of Yes answers in terms of the number of true Yes people. Stanley Warner did this using probability theory, but an alternative is to use simulation, either physical simulation with actual coins or by computer, to study the relationship between the number of Yes answers to a randomized response survey and the actual make-up of the group taking the survey.

6.4 Questions to explore

To explore the workings of the randomized response procedure, you will need a computer program that can simulate a large number of repetitions of the procedure. The first program for this purpose is RAN1RESP.

The program RAN1RESP

This program simulates a randomized response survey. The program requires the following information.

The group:

- Group size: (for example, 20)
- Number of Yes in the group: (any number between 0 and the group size)

The survey:

- Probability of the real question = chance the dime lands Heads: (0.5 for the first set of questions)
- Probability of a Yes answer to the decoy question = chance the penny land Heads: (also 0.5 for the first set of questions)

The simulations:

- Number of replications: (Try 10 the first time you run the program, and 100 the second time, just to get a feel for how the program works. After that, using 1000 replications will usually give good enough results.)
- How often you want to see the results: (The first time, type 1 to see the results after every simulation. The second time, you might type 10, to see the results after every 10th. After that, if you don't want to see any intermediate results.)

Because this program is complex, we don't include the pseudocode for it. However, we have been generous with comments for the human reader in the *MATLAB/Octave* code at the end of the chapter, so you should try to read that and see whether you can follow the logic of the program. We have also included *Mathematica* code at the end of the chapter, but because of the nature of the language it is harder to follow.

Key variables used in RAN1RESP Program

GrpSize	=	group size
YesInGrp	=	number of true Yes in the group
PrRealQ	=	probability of the real question (= probability that the dime lands Heads)
PrYesDecoy	=	probability of Yes to the decoy question (= probability that the penny lands Heads)
NRep	=	number of replications of the survey
PrntFreq	=	how often you see the results of the simulation
NYes	=	number of Yes answers to the survey

6.4.1 Finding estimates

Question 1: *Finding an estimate.* Fix the class size at 20. Let p stand for the proportion of true Yes answers in the class:

$$p = \text{YesInGrp}/20.$$

Use RAN1RESP to study the relationship between the value of p and the expected value (long run average) of the proportion $Y/20$ of Yes answers to the survey:

$$Y/20 = \text{NYes}/20.$$

Your goal should be to find the graph of a function f for which

$$\text{EV}(Y/20) = f(p).$$

Once you have the graph, try to decide the equation of the function, and then use the equation, together with the general principle of Section 6.3, to convert the observed number of Yes answers into an estimate of p:

$$Y/20 \approx \text{EV}(Y/20)$$

by the general principle, and so

$$Y/20 \approx f(p),$$

which gives an estimate of

$$\hat{p} = f^{-1}(Y/20).$$

Note that we use the symbol $\hat{p}$ to represent our estimate of p.

Question 2: *Generalization.* Would you expect the function f to be the same, or different, for a different class size N? With class size N the proportion p of true Yes people and the proportion Y/N of people answering Yes to the survey are given by

$$p = \frac{\text{YesInGrp}}{\text{GrpSize}} \quad \text{and} \quad \frac{Y}{N} = \frac{\text{NYes}}{\text{GrpSize}}.$$

Repeat Question 1, with enough different class sizes that you can find a function $g(p; N)$ such that

$$\text{EV}(Y/N) = g(p; N).$$

Then use the same logic as before to find the estimate $\hat{p}$. Suggestion: Take some time to plan your choices of N and p before you begin your simulations. Then, as you accumulate results, think about what they tell you, and what you still want to find out. Thoughtful choices for N and p can save you time.

Question 3: *Theoretical justification.* Find an informal proof for the result in Question 2. (The form of the output from the program RAN1RESP is designed to help you think about this question. You might want to run the program again, and look at the various relationships among the numbers of Yes and No answers in the three two-way tables.) You will probably need the following results for your argument.

- *Expected values add:*

  $\text{EV}(A + B) = \text{EV}(A) + \text{EV}(B)$

- *Expected number of Heads:*

  For a set of tosses with constant probability of getting Heads (denoted $Pr(\text{Heads})$),

  $\text{EV}(\#\ \text{Heads}) = (\#\ \text{tosses})Pr(\text{Heads})$

6.4.2 Basic properties of the estimate

You will need a second program, RAN2RESP to answer the questions that follow.

The program RAN2RESP: Random

Overview. This program lets you define an estimate (based on the number of Yes answers to a randomized response survey) and investigate various features that measure how well it performs in a long sequence of repeated uses. For example, you might define the estimate to be the one from your answer to Question 2 and define the feature of interest to be the absolute value of the difference between the estimate and the true value. Running the program would then give you the average value of this absolute difference, averaged over repetitions of the survey.

Input. When you run the program, the computer will ask you for exactly the same information as for RAN1RESP. However, before you can run the program, you must first define the estimate you want to study, and the numerical property whose average value you want to compute by simulation.

Defining the estimate. Here is how the current *MATLAB/Octave* version of the program defines the function Estimate:

```
Estimate = NYes/GrpSize;
```

and in *Mathematica*:

Estimate[GrpSize_, NYes_, PrRealQ_, PrYesDecoy_]:=NYes/GrpSize;

As it stands, the function computes the observed fraction of Yes answers (Y/N) from the survey. (This is not a good estimate of $p =$ the fraction of true Yes people in the group.) You will need to replace the above line of code with one or more lines defining the estimate you want to study.

Defining the operating characteristic. The function OpChar(Est, True) defines the particular measure of performance you want to study. (Such measures are known in statistics as operating characteristics.) The variable True is the proportion p of true Yes people in the group:

$$\text{True} = \frac{\text{YesInGrp}}{\text{GrpSize}} = p$$

As written, the code simply gives the value of the estimate itself. In *MATLAB/Octave*:

```
OpChar = Estimate;
```

In *Mathematica*:

```
OpChar[Estimate_, True_]:=Estimate;
```

Running the program will then give you the average value of the estimate, averaged over repetitions of the survey. To study some other feature of the estimate, you will need to replace the definition of OpChar with one of more lines of your own. The various definitions you will need to answer the questions in this section are provided as required.

Question 4: *Expected value of the estimate and bias.*

The **bias** of an estimate is the difference between its expected value and the true value to be estimated. For the randomized response survey, the bias of the estimated proportion of true Yes answers is thus

$$\text{Bias} = b(p, N) = \text{EV}(\hat{p}) - p.$$

Recall that a good estimate of $\text{EV}(\hat{p})$ is the average of the values of $\hat{p}$ obtained over a large number of repeats (say 1000). Notice that because expected values add, $\text{EV}(\hat{p}) - p = \text{EV}(\hat{p} - p)$. We say an estimate is **unbiased** if the bias is equal to zero for all values of the true value being estimated.

The current version of the program RAN2RESP defines Estimate = NYes/GrpSize. This would be a good choice if everyone answered the real question (truthfully) and no one answered the decoy. However, it is unlikely to happen just like that, and NYes/GrpSize is not such a good estimate—among other things, it is biased. Try to find the function $b(p, N)$ that gives the bias in terms of class size N and proportion p of true Yes people in the class. To answer this question you will have to modify the definition of OpChar in RAN2RESP, using the line in *MATLAB/Octave*:

```
OpChar = Estimate - True;
```

or the line in *Mathematica*:

```
OpChar[Estimate_, True_]:=Estimate - True;
```

Question 5: With your results for Estimate $= Y/N =$ NYes/GrpSize as a standard for comparison, change the function Estimate in RAN2RESP so that it uses your randomized response estimate from Section 6.4.1:

$$\text{Estimate} = f^{-1}(Y/N).$$

Compare the results you get now with what you got for Estimate $= Y/N$. Which estimate has greater bias?

Note: For the remaining questions in Section 6.4.2, use your estimate from Question 5.

Question 6: *Typical error size for the estimate.* The bias measures the difference between an estimate's EV and the true value to be estimated. We can also ask about the typical size of the difference between the estimate and its expected value. Here are two ways to measure the typical size:

Mean Absolute Deviation:

MAD $=$ average value of $|\text{estimate} - \text{EV(estimate)}|$.

Standard Deviation:

The SD is similar to the MAD, but instead of taking the *absolute value* of the error, you take the *square* of the error, and then after you take the average, you take the square root, to undo the squaring:

SD $= \sqrt{\text{average value of } [\text{estimate} - EV(\text{estimate})]^2}$

Use RAN2RESP to investigate the relationship between the typical error size of the estimate and the size N of the class, for various choices of p.

(a) Fix the class size N at 20, and regard MAD and SD as functions of p. How would you describe these functional relationships?

(b) Suppose you had used a different class size. Would the functional relationships in (a) be different? Pick a different N, and gather data to see if your guess is supported.

(c) Now fix the value of p at 0.5 and investigate MAD and SD as functions of N.

Suggestion: Think in terms of graphing MAD or SD as a function of N. It will turn out that SD is roughly equal to a power of N times a quantity $h(p)$ that doesn't depend on N:

$$\text{SD} \approx h(p)N^k$$

Thus, for fixed values of p, the SD on the left hand side depends only on N, and is roughly of the form

$$\text{SD} \approx (\text{const1})N^{\text{const2}}$$

Taking logarithms of both sides gives

$$\log(\text{SD}) = \log(\text{const1}) + (\text{const2}) \log N.$$

If you regard log(SD) as y and $\log N$ as x, this is the equation of a line:

$$y = \text{intercept} + (\text{slope})x$$

The form of the equation suggests that if you use the simulation program RAN2RESP to get values of SD for various choices of N, and then plot number pairs $(x, y) = (\log N, \log \text{SD})$, the resulting points should lie near a line with slope equal to const2 and intercept equal to log(const1).

Note that if you plan to plot SD versus $\log(N)$, it would be a good strategy to choose values of N that are equally spaced on a logarithmic scale, for example, 10, 20, 40, 80, 160.

Both MAD and SD are much simpler to compute for estimators that are unbiased. (See Question 4.) For unbiased estimators, the expected value of the estimate equals the true value to be estimated, and so you can use the true value in the definition of OpChar, instead of first having to find the expected value. Here are the definitions to use for unbiased estimators:

For MAD in *MATLAB/Octave*,

```
OpChar = abs(Estimate - True);
```

and in *Mathematica*:

```
OpChar[Estimate_, True_]:=Abs[Estimate − True];
```

For SD in *MATLAB/Octave*,

```
OpChar = (Estimate - True)^2;
```

and in *Mathematica*:

```
OpChar[Estimate_, True_]:=(Estimate − True)^2;
```

to get the average value of the square of $(\text{Est} - \text{True})^2$. Then take the square root to get the SD.

If your estimate is biased, you must run the program two times ((a) and (b) below), first to find the expected value (long run average value) of the estimate itself, and then a second time to find the absolute or squared distance from the estimate to its expected value.

(a) Find the expected value of the estimate. To do this, run the program using

```
OpChar = Estimate;
```

or:

OpChar[Estimate_, True_]:=Estimate;

and take the resulting average value of OpChar as the EV.

(b) Find the average value of the absolute or squared distance from the estimate to its EV. First redefine OpChar using the numerical value of the EV from (a) in place of True in the definitions above. For example, suppose you find in (a) that the expected value for your estimate is 0.4. For MAD,

```
OpChar = abs(Estimate - 0.4);
```

or:

OpChar[Estimate_, True_]:=Abs[Estimate − 0.4];

and run the program to get the average value of this absolute difference. For SD,

```
OpChar = (Estimate - 0.4)^2;
```

or:

OpChar[Estimate_, True_]:=(Estimate − 0.4)^2;

and run the program to find the average value of $(\text{Estimate}-0.4)^2$. Then take the square root of the result.

6.4.3 Changing the chances

If you conduct a randomized response survey using fair coins, then for both the penny and the dime the chance of heads is 1/2. If you use spinners instead of coins, you can set the chances to whatever values you like (Figure 6.1).

The reason for using chances other than 1/2 for either the penny or the dime is to improve the estimate for p, by reducing its SD, for example.

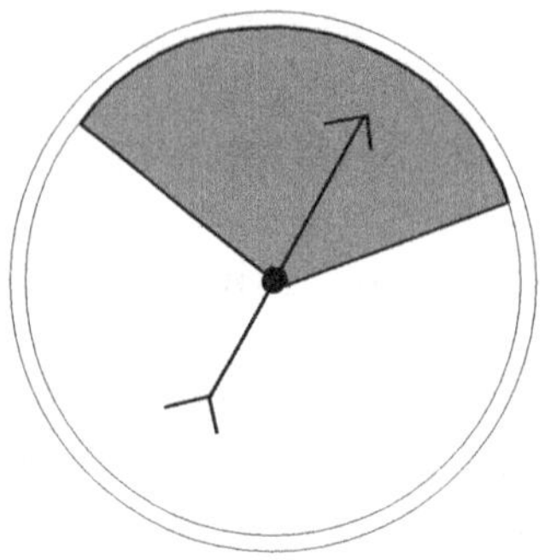

Fig. 6.1: Spinner

Changing Pr(Heads) for the penny (Questions 7-9)

Question 7: Without doing any simulations, guess the general shape of the functional relation between Pr(Heads) for the penny and the SD of the estimate. (Assume that the dime is an ordinary fair coin.) How do you think the estimator will behave if Pr(Heads) is near 0? near 1? Record your guess in the form of a sketch of a graph of SD(θ) as a function of $\theta = Pr$(Heads).

Question 8:. Use the methods of Questions 1–3 from Section 6.4.1 to find a formula for the estimate for p (the true fraction of Yes in the class) in terms of $Y = \#$ Yes answers to the survey, $N =$ size of the class, and $\theta = Pr$(Heads) for the penny.

Question 9: Now hold N fixed at 20, hold p fixed at 0.3, and choose a range of values for θ. For each choice of θ, use the simulation program to compute the SD of the resulting estimate. Graph the set of number pairs (θ, SD), and compare the pattern with your guess in Question 1 above. Write a sentence or two explaining why the relationship between Pr(Heads) and SD has the shape that it does.

Changing Pr(Heads) for the dime (Questions 10-11)

Question 10: Now assume the penny is once again an ordinary fair coin, and let the chance of heads for the dime be some number other than 1/2. Without doing any simulations, guess the relation between Pr(Heads) for the dime and the SD of the estimate. Record your guess in the form of a sketch of a graph of SD(ϕ) as a function of $\phi = Pr$(Heads).

Question 11: Find a formula for the estimate for p in terms of $Y = \#$ Yes answers to the survey, N = size of the class, and $\phi = Pr(\text{Heads})$ for the dime.

Question 12: Now hold N fixed at 20, hold p fixed at 0.3, and choose a range of values for ϕ. For each choice of ϕ, use the simulation program to compute the SD of the resulting estimate. Graph the set of number pairs (ϕ, SD), and compare the pattern with your guess in Question 1 above. Write a sentence or two explaining why the relationship between $Pr(\text{Heads})$ and SD has the shape that it does.

Question 13: Which has a greater effect on the SD of the estimate, $Pr(\text{Heads})$ for the penny, or $Pr(\text{Heads})$ for the dime? If you were designing your own randomized response survey, what values of $\theta = Pr(\text{Heads})$ for the penny and $\phi = Pr(\text{Heads})$ for the dime would you use? Give reasons for your choices.

6.4.4 Other properties of estimates

In Section 6.4.2, you used expected values and standard deviations to evaluate the randomized response estimator. This section begins by introducing two other important features you can use to compare estimators—consistency (Question 14) and large sample normality (Question 15). Both of these ideas are prominent in the mathematical theory of statistics.

Next, Question 16 deals with an unfortunate shortcoming of the randomized response estimator—it can give meaningless values as estimates. Question 17 invites you to study the properties of an improved estimator. Finally, Question 18 introduces another important criterion, mean squared error, for evaluating estimators.

Question 14: *The chance of an error of given size.* For fixed choice of class size N, true proportion p, and a small positive number ϵ, use RAN2RESP to find a numerical value of $Pr(|\hat{p} - p| > \epsilon)$. Now carefully choose a collection of values for N, p, and ϵ, and use these to investigate $Pr(|\hat{p} - p| > \epsilon)$ as a function of N, p, and ϵ. You will have to modify OpChar again. This time, for $\epsilon = 0.05$, say, you would use

```
OpChar=abs(Estimate - True) > 0.05
```

or

```
OpChar[Estimate_, TrueYes_]:=Abs[(Estimate − TrueYes)] > 0.05/.
{False->0, True->1}
```

This code will yield a value of 1 if the error is greater than 0.05 and 0 otherwise. The sum of the 1's and 0's will tell you the number of times the error was more than 0.05, and the average (= Sum / NReps) will tell the proportion of times.

Use the same strategy as in Section 6.4.2 with the SD: is there a power k for which

$$Pr(|\hat{p} - p| > \epsilon) \approx h(p, \epsilon)N^k\,?$$

An estimator is **consistent** if $Pr(|\hat{p} - p| > \epsilon) \to 0$ as $N \to \infty$ for every positive ϵ. Is the randomized response estimator consistent?

Question 15: *Normal approximation.* According to mathematical theory (which says that as $N \to \infty$, deviations tend to follow a bell-shaped normal curve) as the class size N increases, the limit of the chance that an estimate is within 1 SD of its true value is approximately 0.68; the chance that it is within 2 SDs of the true value is 0.95, and the chance that it is within 3 SDs of the true value is 0.997.

The theory says the approximation works well if N is large enough, but it doesn't tell how large N needs to be. Before cheap computers, practicing statisticians would often rely on approximations like this one, even though they had no good way to check how well it worked.

Use different values of N and p, and your SDs from Question 6, to answer the following: How big does N have to be if you want

$$Pr(|\hat{p} - p| \leq 1 \text{ SD }) \approx 0.68$$

to be a good approximation for all choices of p?

Question 16: *The chance of a meaningless value.* Probabilities, by definition, must be non-negative and no greater than one. Values less than zero or greater than one are meaningless. Nevertheless, it is possible to get randomized response estimates that are less than zero or greater than one. For given values of p and N, there is not a simple formula that gives an exact answer for the chance of an estimate less than zero or greater than one, but try to see how well you can describe the general patterns.

(a) For example, fix the class size (75 or 100 is a good choice) and study the relationship between the chance of a meaningless value for the estimate and the true value of p for the class. Are meaningless values more likely when p is near 0? 0.5? 1? Sketch a graph of $Pr(\text{meaningless value})$ versus p. What can you say about symmetry, concavity, etc.?

(b) Now hold p fixed and study the chance that the estimate is not in $[0, 1]$, i.e., $Pr(\hat{p} < 0 \quad \text{or} \quad \hat{p} > 1)$, as a function of N. Here, as often, it can help to start with extreme cases. What can you say if $N = 1$? $N = 2$? What is the limit as N goes to infinity?

This time you want OpChar to return a value of 0 if the estimate is in the interval [0, 1] and return a value of 1 otherwise. Then the average over repetitions will give the proportion of times the value was outside the interval of meaningful values. Here's one way to do this:

```
OpChar = Estimate < 0 | Estimate > 1 ;
```

or

```
OpChar[Estimate_, TrueYes_]:=(Estimate < 0||Estimate > 1)/.
{False->0, True->1}
```

Question 17: *An improved estimator?* Here's a simple-minded way to define a new estimator that does not give meaningless values:

$$\tilde{p} = \begin{cases} 0 & \text{if } \hat{p} < 0 \\ \hat{p} & \text{if } 0 \leq \hat{p} \leq 1 \\ 1 & \text{if } \hat{p} > 1 \end{cases}$$

For this question you will need to modify the definition of the estimate. Here are all but one of the lines you will need.

```
Estimate = (this is the line you'll need to supply, based on your
                    answer to Question 2)
if Estimate < 0
  Estimate = 0;
elseif Estimate > 1
  Estimate = 1
end; %if
```

or

Estimate[GrpSize_, NYes_, PrRealQ_, PrYesDecoy_]:=Module[{Est},

Est = XXXXXX;

Est = If[Est < 0, 0, Est]; Est = If[Est > 1, 1, Est]; {Est}]

where **XXXXXX** is the the formula you'll need to supply, based on your answer to Question 2.

Use the modified program to study the following questions:

(a) What can you say about the bias of the new estimator? Is it unbiased for all values of p? Or does the bias depend on p? Fix N, and use various choices of p to investigate the shape of the bias function $\mathrm{EV}(\tilde{p})-p$.

(b) Now fix p, and study the bias $\mathrm{EV}(\tilde{p}) - p$ as N increases. Is the new estimator consistent?

(c) What can you say about the SD of $\tilde{p}$ compared with the SD of $\hat{p}$?

(d) How close can you come to finding the functional form of the SD of the new estimator, as a function of the class size N and the true proportion p?

Question 18: *Mean square error.*

> The **mean square error** for an estimator is
>
> MSE = EV([estimate − true value $]^2$).
>
> It can be proved that the MSE equals the square of the bias plus the square of the SD.

(a) Use the MSE as a measure for comparing the randomized response estimator from Question 2 with the estimator from Question 17. Is one of the two always better (smaller MSE) no matter what the values of N and p, or is one better for some choices of N and p, and the other better for other choices? Try to formulate as general a recommendation as you can about which estimator to use.

(b) Can you find a third estimator whose MSE is smaller than that of either of the estimators in part (a), for all choices of N and p?

6.5 Computer programs

6.5.1 *MATLAB/Octave* programs

Program: RAN1RESP

```
function []=ran1resp(GrpSize,YesInGrp,PrRealQ,PrYesDecoy,NRep,PrntFreq)
%function []=ran1resp(GrpSize,YesInGrp,PrRealQ,PrYesDecoy,NRep,PrntFreq)
%
%Gives count of primes less or equal to given number n
%
%Inputs
% GrpSize - group size
% YesInGrp - number of true "yes" people in group
% PrRealQ - probbility of real question
% PrYesDecoy - probability of "yes" answer to the decoy question
% NRep - number of replications of simulated survey
% PrntFreq - how often to see results of simulation (0 gives only summary)
%Output
% None, summary output tables will be printed to screen
```

```
TrueYes = 1;  %row for true yes respondents
TrueNo = 2;   %row for true no respondents
Real = 1;  %column for real question
Decoy = 2; %column for decoy question
Total = 3; %column or row for totals
NoInGrp = GrpSize - YesInGrp; %number of true no respondents
YesAnsCum = zeros(3); %matrix for cumlative yes answer data across simulations
NoAnsCum = zeros(3);  %matrix for cumlative no answer data across simulations
AllAnsCum = zeros(3);  %matrix for cumlative data across simulations

for k=1:NRep
YesAns = zeros(3); %matrix for yes answer data
NoAns = zeros(3);  %matrix for no answer data

%compute answers for true yes respondents
dimes=rand(1,YesInGrp)<PrRealQ;      %vector of dime flip outcomes 0 or 1
pennies=rand(1,YesInGrp)<PrYesDecoy; %vector of penny flip outcomes 0 or 1
YesAns(TrueYes,Real)=sum(dimes); %all true yes answer yes to real question
YesAns(TrueYes,Decoy)=sum((1-dimes).*pennies); %got decoy and answered yes
NoAns(TrueYes,Decoy)=sum((1-dimes).*(1-pennies)); %got decoy and answered no

%compute answers for true no respondents
dimes=rand(1,NoInGrp)<PrRealQ;      %vector of dime flip outcomes 0 or 1
pennies=rand(1,NoInGrp)<PrYesDecoy; %vector of penny flip outcomes 0 or 1
YesAns(TrueNo,Decoy)=sum((1-dimes).*pennies); %got decoy and answered yes
NoAns(TrueNo,Real)=sum(dimes);     %all true no answer no to real question
NoAns(TrueNo,Decoy)=sum((1-dimes).*(1-pennies)); %got decoy and answered no

%compute totals
YesAns(TrueYes,Total)=YesAns(TrueYes,Real)+YesAns(TrueYes,Decoy);
YesAns(TrueNo,Total)=YesAns(TrueNo,Real)+YesAns(TrueNo,Decoy);
%total row sum of other rows
YesAns(Total,:)=YesAns(TrueYes,:)+YesAns(TrueNo,:);

NoAns(TrueYes,Total)=NoAns(TrueYes,Real)+NoAns(TrueYes,Decoy);
NoAns(TrueNo,Total)=NoAns(TrueNo,Real)+NoAns(TrueNo,Decoy);
%total row sum of other rows
NoAns(Total,:)=NoAns(TrueYes,:)+NoAns(TrueNo,:);

AllAns = YesAns+NoAns; %Summary for all respondents

%Update Cumaltive Data
YesAnsCum=YesAnsCum+YesAns;
NoAnsCum=NoAnsCum+NoAns;
AllAnsCum=AllAnsCum+AllAns;

if PrntFreq>0 && mod(k,PrntFreq)==0 %check to print
fprintf('==============================================\n')
fprintf('\t     Simulation Number %u \n',k)
fprintf('==============================================\n')

fprintf('\nYES ANSWERS\n')
fprintf('\n\t \t Real  \t   Decoy  \t Total \n')
fprintf('True Yes  %10u  %10u %12u \n',YesAns(1,1),YesAns(1,2),YesAns(1,3))
fprintf('True No   %10u  %10u %12u \n',YesAns(2,1),YesAns(2,2),YesAns(2,3))
fprintf('Total     %10u  %10u %12u \n',YesAns(3,1),YesAns(3,2),YesAns(3,3))
```

```
fprintf('-----------------------------------------------\n\n')


fprintf('NO ANSWERS\n')
fprintf('\n\t \t Real  \t   Decoy  \t Total \n')
fprintf('True Yes  %10u  %10u %12u \n',NoAns(1,1),NoAns(1,2),NoAns(1,3))
fprintf('True No   %10u  %10u %12u \n',NoAns(2,1),NoAns(2,2),NoAns(2,3))
fprintf('Total     %10u  %10u %12u \n',NoAns(3,1),NoAns(3,2),NoAns(3,3))
fprintf('-----------------------------------------------\n\n')

fprintf('ALL ANSWERS\n')
fprintf('\n\t \t Real  \t   Decoy  \t Total \n')
fprintf('True Yes  %10u  %10u %12u \n',AllAns(1,1),AllAns(1,2),AllAns(1,3))
fprintf('True No   %10u  %10u %12u \n',AllAns(2,1),AllAns(2,2),AllAns(2,3))
fprintf('Total     %10u  %10u %12u \n',AllAns(3,1),AllAns(3,2),AllAns(3,3))
fprintf('-----------------------------------------------\n\n')

fprintf('AVERAGES BASED ON %u SIMULATIONS\n',k)
fprintf('------------------------------------\n')

printf('\nAVERAGE YES ANSWERS\n')
fprintf('\n\t \tReal  \t   Decoy  \tTotal \n')
fprintf('True Yes  %10.2f  %10.2f %12.2f \n',YesAnsCum(1,1)/k,YesAnsCum(1,2)/k,YesAnsCum(1,3)/k)
fprintf('True No   %10.2f  %10.2f %12.2f \n',YesAnsCum(2,1)/k,YesAnsCum(2,2)/k,YesAnsCum(2,3)/k)
fprintf('Total     %10.2f  %10.2f %12.2f \n',YesAnsCum(3,1)/k,YesAnsCum(3,2)/k,YesAnsCum(3,3)/k)
fprintf('-----------------------------------------------\n\n')


fprintf('AVERAGE NO ANSWERS\n')
fprintf('\n\t \tReal  \t   Decoy  \tTotal \n')
fprintf('True Yes  %10.2f  %10.2f %12.2f \n',NoAnsCum(1,1)/k,NoAnsCum(1,2)/k,NoAnsCum(1,3)/k)
fprintf('True No   %10.2f  %10.2f %12.2f \n',NoAnsCum(2,1)/k,NoAnsCum(2,2)/k,NoAnsCum(2,3)/k)
fprintf('Total     %10.2f  %10.2f %12.2f \n',NoAnsCum(3,1)/k,NoAnsCum(3,2)/k,NoAnsCum(3,3)/k)
fprintf('-----------------------------------------------\n\n')

fprintf('AVERAGE ALL ANSWERS\n')
fprintf('\n\t \tReal  \t   Decoy  \tTotal \n')
fprintf('True Yes  %10.2f  %10.2f %12.2f \n',AllAnsCum(1,1)/k,AllAnsCum(1,2)/k,AllAnsCum(1,3)/k)
fprintf('True No   %10.2f  %10.2f %12.2f \n',AllAnsCum(2,1)/k,AllAnsCum(2,2)/k,AllAnsCum(2,3)/k)
fprintf('Total     %10.2f  %10.2f %12.2f \n\n',AllAnsCum(3,1)/k,AllAnsCum(3,2)/k,AllAnsCum(3,3)/k)
end; %if
end; %for

if PrntFreq=0 | mod(k,PrntFreq)>0 %if final result not printed, print it
fprintf('\n------------------------------------\n')
fprintf('AVERAGES BASED ON %u SIMULATIONS\n',k)
fprintf('------------------------------------\n')

fprintf('\nAVERAGE YES ANSWERS\n')
fprintf('\n\t \tReal  \t   Decoy  \tTotal \n')
fprintf('True Yes  %10.2f  %10.2f %12.2f \n',YesAnsCum(1,1)/k,YesAnsCum(1,2)/k,YesAnsCum(1,3)/k)
fprintf('True No   %10.2f  %10.2f %12.2f \n',YesAnsCum(2,1)/k,YesAnsCum(2,2)/k,YesAnsCum(2,3)/k)
fprintf('Total     %10.2f  %10.2f %12.2f \n',YesAnsCum(3,1)/k,YesAnsCum(3,2)/k,YesAnsCum(3,3)/k)
fprintf('-----------------------------------------------\n\n')


fprintf('AVERAGE NO ANSWERS\n')
```

```
fprintf('\n\t \tReal  \t   Decoy  \tTotal \n')
fprintf('True Yes  %10.2f  %10.2f %12.2f \n',NoAnsCum(1,1)/k,NoAnsCum(1,2)/k,NoAnsCum(1,3)/k)
fprintf('True No   %10.2f  %10.2f %12.2f \n',NoAnsCum(2,1)/k,NoAnsCum(2,2)/k,NoAnsCum(2,3)/k)
fprintf('Total     %10.2f  %10.2f %12.2f \n',NoAnsCum(3,1)/k,NoAnsCum(3,2)/k,NoAnsCum(3,3)/k)
fprintf('---------------------------------------------\n\n')

fprintf('AVERAGE ALL ANSWERS\n')
fprintf('\n\t \tReal  \t   Decoy  \tTotal \n')
fprintf('True Yes  %10.2f  %10.2f %12.2f \n',AllAnsCum(1,1)/k,AllAnsCum(1,2)/k,AllAnsCum(1,3)/k)
fprintf('True No   %10.2f  %10.2f %12.2f \n',AllAnsCum(2,1)/k,AllAnsCum(2,2)/k,AllAnsCum(2,3)/k)
fprintf('Total     %10.2f  %10.2f %12.2f \n',AllAnsCum(3,1)/k,AllAnsCum(3,2)/k,AllAnsCum(3,3)/k)
fprintf('---------------------------------------------\n\n')
end; %if
```

Program: RAN2RESP

```
function OpCharAvg=ran2resp(GrpSize,YesInGrp,PrRealQ,PrYesDecoy,NRep,PrntFreq)
%function OpCharAvg=ran2resp(GrpSize,YesInGrp,PrRealQ,PrYesDecoy,NRep,PrntFreq)
%
%Gives count of primes less or equal to given number n
%
%Inputs
%  GrpSize - group size
%  YesInGrp - number of true "yes" people in group
%  PrRealQ - probbility of real question
%  PrYesDecoy - probability of "yes" answer to the decoy question
%  NRep - number of replications of simulated survey
%  PrntFreq - how often to see results of simulation (0 = only final result)
%Output
%  OpCharAvg, summary output tables will be printed to screen

OpCharAvg=0;  %Variable to track average operating characteristic
NoInGrp=GrpSize-YesInGrp; %number of True No respondents
True = YesInGrp/GrpSize; %true proportion of Yes in group

for k=1:NRep
%compute answers for true yes respondents
dimes=rand(1,YesInGrp)<PrRealQ;       %vector of dime flip outcomes 0 or 1
pennies=rand(1,YesInGrp)<PrYesDecoy; %vector of penny flip outcomes 0 or 1
NYes=sum(dimes); %all true yes answer yes to real question
NYes=NYes+sum((1-dimes).*pennies); %got decoy and answered yes

%compute answers for true no respondents
dimes=rand(1,NoInGrp)<PrRealQ;       %vector of dime flip outcomes 0 or 1
pennies=rand(1,NoInGrp)<PrYesDecoy; %vector of penny flip outcomes 0 or 1
NYes=NYes+sum((1-dimes).*pennies); %got decoy and answered yes

%************ SET ESTIMATE AND OPCHAR HERE ************%
Estimate = NYes/GrpSize;
OpChar = Estimate;
```

```
%****************************************************%

%Update Cumaltive Data
OpCharAvg=OpCharAvg+OpChar;

if PrntFreq>0 && mod(k,PrntFreq)==0 %check to print
fprintf('Average Operating Characteristic after %u steps is %g \n\n',k,OpCharAvg/k)
end; %if
end; %for
OpCharAvg=OpCharAvg/NRep;
```

6.5.2 *Mathematica* programs

Survey Simulation

The following program, Ran1Resp, simulates a randomized response survey. It takes 5 paramaters; in order, they are: GrpSize, the size of the group being surveyed; YesInGrp, the number of people in the group whose true answer to the question is Yes (in the chapter, p is YesInGrp divided by N); PrRealQ, the probability that a person will answer the real question (the probability that the dime lands heads); PrYesDecoy, the probability that the answer to the decoy question is yes (in the chapter, the probability that the penny lands heads); and NRep, the number of surveys the program will simulate.

```
Ran1Resp[GrpSize_, YesInGrp_, PrRealQ_, PrYesDecoy_, NRep_]:=
  Module[{Running = Result = ZeTab = Table[0, {3}, {3}, {3}], i, j, TrueYes = 1,
    TrueNo = 2, Real = 1, Decoy = 2, Yes = 1, No = 2, Total = 3,
    topper = {" YES ", " NO ", "TOTAL"}}, Do[Result = ZeTab;
    Do[If[RandomReal[] < PrRealQ, Result[[TrueYes, Real, Yes]]++,
      If[RandomReal[] < PrYesDecoy, Result[[TrueYes, Decoy, Yes]]++,
        Result[[TrueYes, Decoy, No]]++]], {YesInGrp}];
    Do[If[RandomReal[] < PrRealQ, Result[[TrueNo, Real, No]]++,
      If[RandomReal[] < PrYesDecoy, Result[[TrueNo, Decoy, Yes]]++,
        Result[[TrueNo, Decoy, No]]++]], {GrpSize − YesInGrp}];
```

```
    Do[Result[[i, j, Total]] = Result[[i, j, Yes]] + Result[[i, j, No]];
      Result[[i, Total, j]] = Result[[i, Real, j]] + Result[[i, Decoy, j]];
      Result[[Total, i, j]] = Result[[TrueYes, i, j]] + Result[[TrueNo, i, j]], {i, 3}, {j, 3}];
    Running = Running + Result, {NRep}];
If[NRep > 1, Running = Round[Running 10^1 / NRep] / 10^1];
Print["        T R U E  Y E S       T R U E  N O     W H O L E  G R O U P"];
Print[TableForm[{" ", "REAL", "DECOY", "TOTAL"}], "\t",
    TableForm[Prepend[Running[[1]], topper]],
    "\t", TableForm[Prepend[Running[[2]], topper]], "\t",
    TableForm[Prepend[Running[[3]], topper]]]]
```

Here is an example (maximizing the screen will make the display easier to read).In this example, the group size is 17, the number of people whose true answer to the question is yes is 12, the probabilities of the penny and dime landing head are each .5, and the simulation will run 20 times.

```
Ran1Resp[17, 12, .5, .5, 20]
```

Properties of the Estimate

The following program, Ran2Resp, will help us explore some properties of the estimate you are using. Be sure to also define Estimate and OpChar (below) before running Ran2Resp!

```
Ran2Resp[GrpSize_, YesInGrp_, PrRealQ_, PrYesDecoy_, NRep_]:=
  Module[{RunningSum = 0, NYes}, Do[NYes = 0;
    Do[If[RandomReal[] < PrRealQ||RandomReal[] <
      PrYesDecoy, NYes++], {YesInGrp}];
    Do[If[!RandomReal[] < PrRealQ&&RandomReal[] <
```

```
        PrYesDecoy, NYes++], {GrpSize - YesInGrp}];
        RunningSum = RunningSum+
OpChar [Estimate[GrpSize, NYes, PrRealQ, PrYesDecoy], YesInGrp/GrpSize], {NRep}] ;
N[RunningSum/NRep]] ;
```

To study properties of your estimate of the proportion of "True Yesses" (people whose true answer to the question is Yes, i.e., YesInGrp divided by GrpSize), set Estimate to your estimate. (In the chapter, the true proportion is denoted by p, and your estimate is denoted by $\hat{p}$.) When you open this notebook the first time, Estimate is set to NYes divided by GrpSize, which is simply assuming that the number of True Yesses is simply the number of people reporting a Yes after flipping coins. So be sure to reset it to your estimate.

```
Estimate[GrpSize_, NYes_, PrRealQ_, PrYesDecoy_]:=NYes/GrpSize;
```

OpChar is short for "operating characteristic". You change its definition, depending on which operating characteristic of your estimate you want to investigate. To study bias, set it to Estimate - True; to study mean absolute deviation (MAD), set it to Abs[Estimate - True]; to study standard deviation, set it to (Estimate - True)^2.

```
OpChar[Estimate_, True_]:=Estimate;
```

Now that we have set OpChar and Estimate, we can run Ran2Resp

```
Ran2Resp[17, 12, .5, .5, 20]
```

Chapter 7

Polyhedra

7.1 Introduction

In this chapter you will look at (*really* look at—you'll be making models) three dimensional shapes of various kinds and investigate some of their properties. Unlike other chapters, this one makes no use of the computer. Answering the questions will require you to model, look and think.

A *regular polygon* is a plane figure, all of whose sides and interior angles are equal. There exist regular polygons with any number of sides (greater than 3) and any size.

Question 1: If a polygon has n sides, how many vertices does it have?

Question 2: Find a formula for the area of a polygon with n sides, each of which has length 1.

A polyhedron is a solid whose all of whose faces are polygons. Thus, a tetrahedron, a rectangular box, a pyramid, and a truncated pyramid are all polyhedra. A *regular polyhedron* (or a *regular solid*) is one in which all faces are congruent polygons, and the same number of polygons meet around each vertex.

7.2 Questions and discussion

Question 3: Cut out a large supply of triangles, squares, pentagons, hexagons, septagons, octagons, ...

(a) Can you make a regular solid out of (equilateral) triangles? If so, do it. If not, why not?

(b) Can you make a regular solid out of squares? If so, do it. If not, why not?

(c) Can you make a regular solid out of regular pentagons? If so, do it. If not, why not?

(d) Can you make a regular solid out of regular hexagons? If so, do it. If not, why not?

(e) Can you make a regular solid out of regular heptagons? If so, do it. If not, why not?

(f) Can you make a regular solid out of regular octagons? If so, do it. If not, why not?

(g) Guess the next question, and the next, and ... Answer them.

Question 4: The Greeks said that there were only five regular solids. Find five different regular solids and describe them carefully. In particular, how many faces does each have? How many vertices? How many edges? Were the Greeks right? Proof or counterexample.

Question 5: Find the diameter of each of the regular solids if the edge length is 1. What is the diameter of each of the regular solids if the faces have area 1?

One could also look for the "simplest" polyhedra, with some suitable notion of simple. For example, we might look for those with the smallest number of vertices. Any polyhedron with just 4 vertices is a tetrahedron (that is, it is bounded by four triangles).

Question 6: Are there different ways to make polyhedra with 5 or 6 vertices? Can you enumerate all polyhedra with at most seven vertices?

A polyhedron is said to be *convex* if, whenever two points lie in the interior of the polyhedron, the segment joining them lies entirely inside the polyhedron. Given any convex polyhedron centered at the origin, we can create a new polyhedron, by "cutting off corners." That is, given a vertex, consider the line joining the center of the polyhedron to the vertex. Starting at the vertex move a short distance inside the polyhedron and slice the polyhedron perpendicular to this line. Do this at every vertex, moving in the same short distance. Here, what we mean by a "short" distance is that no two of the faces made by cutting off the corners share a vertex or edge.

Question 7: Describe the solids you get by cutting off the corners of the regular solids. (i.e., How many faces? What types are they? How many edges? How many vertices?) Is there a way to cut off the corners so the solid looks most symmetrical? Is it possible to arrange that all the faces have area 1? Build at least one model.

Question 8: Describe the solids that you get when you cut off corners of regular solids and let the new faces grow so they just touch one another. What is the relation between the areas of the faces? Build at least one model.

Question 9: Describe a soccer ball. Can you get a soccer ball from one of the regular solids by cutting off corners?

The cover story of the September 1995 issue of the *Notices* of the American Mathematical Society is about a recently discovered molecule called a *Fullerene*—or sometimes a *buckyball*[1] (These molecules are named for Buckminster Fuller.) Quoting from the article by Bertram Kostant,

> Prior to the discovery of the Fullerenes, around ten years ago, the only known form of pure solid carbon was graphite and diamonds. These two forms are crystalline materials where the

[1] "The Graph of the Truncated Icosahedron and the Last Letter of Galois," by Bertram Kostant, *Notices of the AMS*, Volume 42, Number 9, 1995, pages 959-968.

> bonds between the carbon atoms exhibit hexagonal and tetrahedral structures, respectively. In neither of these two substances, however, are there isolated molecules of pure carbon. On the other hand, in Fullerene one finds for the first time a pure carbon crystalline solid with well-defined carbon molecules. .... Mathematically these molecules can be described as convex polyhedrons where the faces are either hexagons or pentagons and each vertex (carbon atom) is the endpoint of three edges (carbon bonds). .... Fullerenes exhibit remarkable chemical and physical properties (e.g., superconductivity, ferromagnetism, tremendous stability) and have been the objects of a vast amount of research throughout the world.

You can do a little research on buckyballs too.

Question 10: Buckyballs can be constructed by cutting off corners. Describe one. (Hint: Kostant says, "Among the many Fullerene molecules the most prominent and the most studied is C_{60}. Can you describe C_{60}?)

Question 11: Pick your favorite regular solid. Describe what happens when you iterate the procedure of cutting off corners. Answer the question using both the procedure in Question 6 and that in Question 7.

Question 12: On answering Questions 1-10, you should have accumulated lots of data regarding polygons with different numbers of edges, vertices, and faces. Is there a relationship among these numbers? What is it? Is it possible to specify the number of vertices and edges independently? What about the number of faces and edges? Vertices and faces? Can you investigate this question completely for, say, number of vertices (or edges, or faces) less than eight? If you can, do it. If not, why not?

7.3 Additional topic

Can you figure out volumes in Questions 7 and 8 under reasonable conditions?

Chapter 8

The p-adic Numbers

8.1 Introduction

This chapter will explore the properties of a new set of numbers called the p-adic numbers. These numbers were first introduced by the German mathematician Kurt Hensel (1861-1941). For the rest of this chapter, let p be any prime number, $p \in \{2, 3, 5, 7, 11, 13, ...\}$. For each distinct prime, there is a new and different set of p-adic numbers. The 2-adics will be denoted by $\mathbb{Q}_2$, the 3-adics by $\mathbb{Q}_3$, the 5-adics by $\mathbb{Q}_5$, and so on. Through this exploration of numbers that are very exotic and unusual, you will come to a better understanding of the familiar real numbers. The real numbers are usually denoted by $\mathbb{R}$ but in this chapter we will use the notation $\mathbb{Q}_\infty$ for the real numbers to emphasize what they have in common with the p-adics. The p-adic numbers have become extremely important in modern number theory, but they are still not well-known to all mathematicians, and they are certainly not a part of the usual undergraduate curriculum. The study of these numbers is exciting because it unites number theory, algebra, analysis, and topology, four different areas of mathematics.

In learning about the p-adic numbers, you will:

- Learn to do arithmetic with these new numbers;

- See these new numbers as the analogues of the of the real numbers and come to see the real numbers in a more abstract way; and
- Get a brief introduction to the topological idea of "nearness" and see that convergence depends upon that idea.

Before we begin a formal introduction to the p-adic numbers, consider the number 0 for a moment. We know that 0 is the only number that is divisible by all other numbers. Suppose we say this by noting that 0 is the only number that is divisible by any prime number p and all its prime powers p^e. We could then begin to think that a number was "close" to 0 if it was divisible by a high power of a prime p. Of course, a number could be close to 0 for one prime and not close for another (625 is "close" to 0 for 5 but not for 3), so we should say that a number is 5-adically "close" to 0 if it is highly divisible by 5. We could now say things like, "For the prime 5, 625 is closer to 0 than 125 is."

Going a bit further with this line of thought, we consider the equation $x - 47 = 0$ and try to find x so that $x - 47$ is "close" to 0 for the prime 5. We can find x so that $x - 47$ is divisible by 5. In the notation of modular arithmetic, we want to solve $x - 47 \equiv 0 \bmod 5$. If you do not know any modular arithmetic, this notation just means that we want to find a number x between 0 and 4 such that $x - 47$ is divisible by 5. Quickly, we see that $x = 2$ works. Next, we ask to find x so that $x - 47$ is divisible by 5^2 and hence "closer" to 0. We want to solve the equation $x - 47 \equiv 0 \bmod 5^2$ and find the number x between 0 and 24 such that $x - 47$ is divisible by 5^2. The answer to this question is $x = 22$. Notice that if $x - 47$ is divisible by 5^2 then it is certainly divisible by 5. This observation explains why $x = 22 = 2 + 4 \cdot 5$. In other words, we extended the solution modulo 5 to a solution modulo 5^2 by adding a multiple of 5. If we ask to solve $x - 47 \equiv 0 \bmod 5^3$, we get $x = 47 = 2 + 4 \cdot 5 + 1 \cdot 5^2$, which is nothing more than the expansion of 47 in powers of 5. If we go on to look at higher powers of 5, our solution for x does not change. This expression for x is also called the 5-adic expansion of 47 in $\mathbb{Q}_5$; it is analogous to the decimal expansion of 47 in $\mathbb{R}$.

If we try this same idea on $x + 47 = 0$, we get something even more intriguing:

The solution to $x + 47 \equiv 0 \bmod 5$ is $x = 3$.

The solution to $x + 47 \equiv 0 \bmod 5^2$ is still $x = 3 = 3 + 0 \cdot 5$.
The solution to $x + 47 \equiv 0 \bmod 5^3$ is $x = 78 = 3 + 0 \cdot 5 + 3 \cdot 5^2$.
(To find 78, we solved the equation $(3 + 0 \cdot 5 + a \cdot 5^2) + 47 \equiv 0 \bmod 5^2$ for a. Try it!)
The solution to $x + 47 \equiv 0 \bmod 5^4$ is $x = 578 = 3 + 0 \cdot 5 + 3 \cdot 5^2 + 4 \cdot 5^3$.
The solution to $x + 47 \equiv 0 \bmod 5^e$ is
$x = 3 + 0 \cdot 5 + 3 \cdot 5^2 + 4 \cdot 5^3 + ... + 4 \cdot 5^{e-1}$.

The infinite expression $x = 3 + 0 \cdot 5 + 3 \cdot 5^2 + 4 \cdot 5^3 + ...$ is the way that -47 or the additive inverse of 47 gets represented in $\mathbb{Q}_5$. What do we get if we add the 5-adic representation of 47 to that of -47? Is it 0? How do we know it must be 0?

Notice that we could continue our process above by finding solutions to equations like $2x - 3 = 0$ (i.e. finding a representation for $2/3$) or $x^2 + 1$ (i.e. finding $\sqrt{-1}$) that are "closer" and "closer" to 0 in this sense. Try it!

In this laboratory, you will begin to figure out what these representations mean for your understanding of mathematics. How can the x above be another way to represent -47? What do you gain by looking at numbers this way? You already have a perfectly good way to represent them.

To begin, we need a more precise idea of what we mean by "closer and closer to 0".

8.2 Absolute values on rational numbers

Let p stand for any fixed prime number $p \in \{2, 3, 5, 7, ...\}$. We can construct the p-adic numbers $\mathbb{Q}_p$ from the rational numbers $\mathbb{Q} = \{a/b$ where $a, b \in$ the integers $\mathbb{Z}\}$, in exactly the same way we construct the real numbers, $\mathbb{Q}_\infty$, from the rational numbers.

An absolute value on the rational numbers is a map, which we will call $|\cdot|$, from $\mathbb{Q}$ to $[0, \infty]$ which has the following properties for any x and y in $\mathbb{Q}$:

1. $|x| \geq 0$, and $|x| = 0$ if and only if $x = 0$.

2. $|x \cdot y| = |x| \cdot |y|$.

3. $|x + y| \leq |x| + |y|$. (The *triangle inequality*)

The German mathematician Ostrowski showed in 1935 that there are only three different types of absolute values that can be put on the rational numbers. All others are equivalent[1] to one of these types. The three types are: the trivial absolute value, the usual absolute value, and the p-adic absolute value:

Definition 1. Let x be a rational number.

The *trivial absolute value* of x, denoted $|x|_0$, is defined by:

$$|x|_0 = \begin{cases} 1 & \text{if } x \neq 0 \\ 0 & \text{if } x = 0. \end{cases}$$

The *usual absolute value* of x, denoted $|x|_\infty$, is defined by:

$$|x|_\infty = \begin{cases} x & \text{if } x \geq 0 \\ -x & \text{if } x < 0. \end{cases}$$

The *p-adic absolute value* of x, denoted $|x|_p$, is defined by:

$$|x|_p = \begin{cases} 1/p^{ord_p(x)} & \text{if } x \neq 0 \\ 0 & \text{if } x = 0. \end{cases}$$

The quantity $ord_p x$ is called the order of x, and it is the highest power of p dividing x. In other words, if $x = p^n \cdot (\alpha/\beta)$, where α and β are integers that are not divisible by p, then $n = ord_p x$. For example, if $p = 5$ then $|75|_5 = |5^2 \cdot 3|_5 = 5^{-2}$, $|34|_5 = |5^0 \cdot 34|_5 = 1$, $\left|\frac{1}{5}\right|_5 = 5$, and $\left|\frac{10}{75}\right|_5 = \left|5^{-1} \cdot \frac{2}{3}\right|_5 = 5$.

Exercise 1: Find the following absolute values:

$$|340|_2, \ |340|_3, \ |340|_5, \ |340|_{17}$$

$$\left|\frac{340}{33}\right|_2, \ \left|\frac{340}{33}\right|_3, \ \left|\frac{340}{33}\right|_5 \ \left|\frac{340}{33}\right|_{19}$$

[1] Two absolute values are equivalent if the same sequences of rational numbers converge for both of them.

Exercise 2: Let $p = 5$. Can you characterize (describe in a sentence) all the rational numbers with 5-adic absolute value equal to 1? Give some examples of these numbers. Make sure that some of your examples are positive, some negative, and that some are fractions. Can you characterize all the rational numbers with 5-adic absolute value equal to $\frac{1}{5}$? Give some examples of these rational numbers. How are these sets related? Given an element of absolute value 1, is there a corresponding element of absolute value $\frac{1}{5}$? What are the other possible values of the 5-adic absolute value? Can you describe the sets of rational numbers that have each possible value? Can you show that 0 is the only rational number of 5-adic absolute value equal to 0?

Exercise 3: Show that each of these absolute values really satisfies the three properties of an absolute value for all rational numbers.

To do this you may want to break the things you have to show into cases. For example, with the trivial absolute value, you want to see what happens when both x and y are 0, when one of them is 0 and the other is not, and when both of them are not 0. With the usual absolute value you will need to consider both when x and y are 0 or not and when they are positive or negative. Additionally, in the p-adic case, you will need to consider when x and y have the same number of p's in their prime factorizations and when they have different numbers of p's.

We say that an absolute value on $\mathbb{Q}$ is *non-Archimedean* if in addition to the three properties above it also satisfies the additional property (sometimes called the ultrametric inequality)

$$\text{3a. } |x + y| \leq \max(|x|, |y|) \text{ for all } x \text{ and } y \in \mathbb{Q}.$$

Property 3a is a stronger property than the triangle inequality because if $|x + y| \leq \max(|x|, |y|)$, then it is certainly less then or equal to the sum of both $|x|$ and $|y|$. Absolute values which do not satisfy property 3a are called *Archimedean* absolute values.[2]

[2] Archimedes was a Greek mathematician who lived from 287-212 B.C. He first formulated the following property, called the Archimedean property:

Given $x, y \in \mathbb{Q}$, where $x \neq 0$,
there exists a positive integer n such that $|nx| > |y|$.

Question 1: Classify each of the three absolute values in Definition 1 as Archimedean or non-Archimedean.

Question 2: Let $p = 5$ and consider the rational number 2. Find $|2|_5$, $|2+2|_5$, $|2+2+2|_5$, $|2+2+2+2|_5$, $|2+2+2+2+2|_5$. What will happen to the absolute values as one adds 2's? Do the sums continue to decrease uniformly in absolute value? Consider the rational number $\frac{2}{25}$. Find $\left|\frac{2}{25}\right|_5$, $\left|\frac{2}{25}+\frac{2}{25}\right|_5$, etc. Describe what happens as one adds more copies of $\frac{2}{25}$.

8.3 The real numbers

We construct the real numbers and the p-adic numbers from the rational numbers by defining them to be the limits of all Cauchy sequences of rational numbers.

Definition 2. A sequence of rational numbers, $\{a_n\} = \{a_1, a_2, a_3, ...\}$, is said to be a *real Cauchy sequence* if given some $\epsilon > 0$, there exists a positive integer N such that for all $i, j > N$, $|a_i - a_j|_\infty < \epsilon$.

Thus, we define the real number $\pi = 3.1415926...$ to be the limit of the sequence of rational numbers that starts off with

$$\begin{aligned} a_1 &= 3, \\ a_2 &= 3.1 = 3 + 1 \cdot 10^{-1}, \\ a_3 &= 3.14 = 3 + 1 \cdot 10^{-1} + 4 \cdot 10^{-2}, \\ &\vdots \end{aligned}$$

This is a Cauchy sequence because if we choose any small positive number epsilon, say take $\epsilon = 10^{-4}$, then we see by our sequence above that we can take $N = 4$ (or, indeed, any integer bigger than 4) and for members of the

Archimedes formulated this property geometrically in terms of line segments. If this property holds then it means that there are integers with arbitrarily large absolute values and that if you add a positive integer to itself again and again the absolute value of the sum will grow without bound.

sequence beyond the fourth member, $i, j > 4$, we will have that $|a_i - a_j|_\infty < \epsilon$.[3] The decimal expansion for π is therefore really short-hand notation for the sequence of rational numbers that converges to π.

Exercise 4: The square root of 2 has a decimal expansion which begins with 1.4142135.... This expansion represents the Cauchy sequences which converge to $\sqrt{2}$; in particular it represents the sequence $\{1, 1.4, 1.41, 1.414, ...\}$. If $\epsilon = 0.05$, find the smallest big N such that $|a_n - a_m|_\infty < .05 = \epsilon$ for $n, m > N$. Next, if $N = 4$ what is the set of all ϵ such that $|a_n - a_m|_\infty < \epsilon$ for $n, m > 4 = N$?

In forming our Cauchy sequences above, the topological idea of an absolute value was key. To construct the real numbers we use the usual absolute value. There is a little ambiguity about the construction, as we have described it, because there are many Cauchy sequences that converge to the same real number. For example, the sequence $\{a_n\} = \{1, 1, 1, 1, ...\}$ converges to 1 but so does the sequence $\{b_n\} = \{1.1, 1.01, 1.001, 1.0001, ...\}$.

We say that two real Cauchy sequences *represent the same real number* (or are *equivalent*) if the sequence $\{a_n - b_n\}$, which in this case is the sequence $\{.1, .01, .001, .0001, ...\}$, converges to 0 (with respect to the usual absolute value). Notice that the sequence $\{b_n\}$ above cannot be represented as a decimal. We use the decimal expansion of a real number to stand for the class of all Cauchy sequences that converge to the same real number. However, even the decimal expansion of a real number is not unique. For example, the sequence $\{c_n\} = \{0.9, 0.99, 0.999, 0.9999, ...\}$, which has decimal expansion $.999\bar{9}$, is another Cauchy sequence representing 1. To see that $.999\bar{9} = 1$, let's sum the infinite series for $0.999\bar{9}$.

$$\begin{aligned} 0.999\bar{9} &= \lim_{n\to\infty} \left(\frac{9}{10} + \frac{9}{100} + \cdots + \frac{9}{10^n} \right) \\ &= \lim_{n\to\infty} \left(\frac{9}{10} \right) (1 + 10^{-1} + 10^{-2} + \cdots + 10^{-(n-1)}). \end{aligned}$$

Since we know that the sum of the geometric series

$$1 + r + r^2 + \cdots + r^{n-1} = \frac{1 - r^n}{1 - r}$$

[3]For example, if $i = 12$ and $j = 5$, we have that $a_{12} = 3.14159265358$ and $a_5 = 3.1415$ and we see that $|a_{12} - a_5|_\infty = 0.00009265358 < 10^{-4}$.

and that $\lim_{n\to\infty} 10^{-n} = 0$, we see that

$$0.999\bar{9} = \left(\frac{9}{10}\right) \lim_{n\to\infty} \left(\frac{1-10^{-n}}{1-10^{-1}}\right) = \left(\frac{9}{10}\right)\left(\frac{1}{1-10^{-1}}\right) = 1.$$

Exercise 5: Sum a geometric series and find the rational number represented by the following repeating decimals: $0.\overline{31}$, $34.\overline{23}$, $123.24\bar{5}$.

Definition 3: The *real numbers* $\mathbb{R}$ are defined to be all equivalence classes of real Cauchy sequences.

Definition 4: A field is *complete* with respect to an absolute value if every Cauchy sequence of numbers in the field converges to a number in the field.

Since the real numbers are defined as limits of Cauchy sequences of rational numbers, it seems clear that the real numbers should form a complete field (a field without holes). There should be no holes in the real numbers since every real number is a Cauchy sequence of rational numbers, and between any two such Cauchy sequences there is always another one. For example, between 3.14159111... and π there are infinitely many other real numbers: 3.141592, 3.1415915, 3.1415925555..., etc. Completeness actually requires that all Cauchy sequences of real numbers (that is, Cauchy sequences of Cauchy sequences) converge to real numbers. We will not prove that the real numbers and the p-adic numbers are complete fields in this chapter, but we just mention this fact to point out that completeness is not obvious.

8.4 The p-adic numbers

To construct the p-adic numbers, we use the p-adic absolute value in the same way we used the usual absolute value above.

A sequence of rational numbers, $\{a_n\} = \{a_1, a_2, a_3, ...\}$, is said to be a *$p$-adic Cauchy sequence* if given some $\epsilon > 0$, there exists a positive integer N such that for all $i, j > N$, $|a_i - a_j|_p < \epsilon$.

Two p-adic Cauchy sequences, $\{a_n\}$ and $\{b_n\}$, are said to be *equivalent* if the sequence $\{a_n - b_n\}$ represents 0. The sequence of differences will represent 0 if $\lim_{n\to\infty} |a_n - b_n|_p = 0$. For example, if $p = 5$ then $\{a_n\} = \{1, 1, 1, ...\}$ and $\{b_n\} = \{1+5, 1+5^2, 1+5^3, ...\}$ are equivalent since $|1-(1+5^n)|_5 = \frac{1}{5^n} \to 0$ as $n \to \infty$.

Definition 5. The *field p-adic numbers* $\mathbb{Q}_p$ is defined to be all equivalence classes of p-adic Cauchy sequences.

Now we know that any p-adic Cauchy sequence must converge to a p-adic number. What do p-adic Cauchy sequences look like, and is there a p-adic expansion for every p-adic Cauchy sequence that we can use to represent the p-adic number just as we use the decimal expansion to represent a real number?

First of all, repeating sequences of positive integers form p-adic Cauchy sequences which converge to the repeated integer. For example, it is certainly true that $\{406, 406, 406, ...\}$ is a Cauchy sequence converging to 406 as $|a_m - a_n|_p = 0 < \epsilon$ for all ϵ and all m and n. If $p = 5$, we can find the 5-adic expansion of 406 by writing the number out in powers of 5 as $406 = 1 + 1 \cdot 5 + 1 \cdot 5^2 + 3 \cdot 5^3$. To write down this expansion, we found that 5^3 was the largest power of 5 still less than or equal to 406. We subtracted off as many of those as we could from 406, leaving us with $406 - 3 \cdot 5^3 = 31$. We then continued the process by subtracting off multiples of 5^2 and 5 from 31 to get the 5-adic expansion. The expansion above will now be our short-hand notation for the class of all Cauchy sequences equivalent to the Cauchy sequence $\{1, 1 + 1 \cdot 5, 1 + 1 \cdot 5 + 1 \cdot 5^2, 1 + 1 \cdot 5 + 1 \cdot 5^2 + 3 \cdot 5^3, ...\}$. If $p = 7$ then $406 = 0 + 2 \cdot 7 + 1 \cdot 7^2 + 1 \cdot 7^3$ will be its 7-adic expansion.

Just as $\{1, 0.1, 0.01, 0.001, ...\}$ (or $\lim_{n\to\infty} 10^{-n}$) is another representation for 0 in the real numbers, $\{5, 5 + 4 \cdot 5, 5 + 4 \cdot 5 + 4 \cdot 5^2, ...\} = \{5, 5^2, 5^3, ...\}$ (or $\lim_{n\to\infty} 5^n$) is another representation for 0 in the 5-adics.

The p-adic expansion of an integer makes it very easy to determine its p-adic absolute value because we can see clearly how many powers of p divide it by looking at the first non-zero term of its p-adic expansion.

Exercise 6: For $p = 5$, find the p-adic expansions, the absolute values, and the ord_p of 700, 34,and 95. Do the same for $p = 7$.

If we go looking for other sequences of integers that are p-adic Cauchy sequences, we see that any sequence of the following form must converge:

$$\{\alpha_n\} = \{a_0,\ a_0 + a_1 \cdot p,\ a_0 + a_1 \cdot p + a_2 \cdot p^2,\ ...\}$$

where the a_j (the coefficients of the powers of p) are integers between 0 and $p-1$. These sequences converge because for $m < n$

$$|\alpha_n - \alpha_m| = |\alpha_{m+1} \cdot p^{m+1} + ... + \alpha_n \cdot p^n|_p \leq p^{-(m+1)}$$

and this $p^{-(m+1)}$ can be made smaller than any given ϵ by taking m and n large enough. We use the p-adic expansion $a_0 + a_1 \cdot p + a_2 \cdot p^2 + a_3 \cdot p^3 + ...$ to stand for the Cauchy sequence above and abbreviate this expansion with the notation $a_0.a_1a_2a_3\,..._p$, just as we do with the decimal expansion for real numbers. For example, we see now that the sequence

$$\{4, 19, 94, 469, ...\} = \{4, 4{+}3{\cdot}5, 4{+}3{\cdot}5{+}3{\cdot}5^2, ...\} = 4{+}3{\cdot}5{+}3{\cdot}5^2{+}... = 4.\bar{3}_5$$

converges in $\mathbb{Q}_5$.

It turns out that it is possible to prove that every p-adic Cauchy sequence of positive integers is always equivalent to one of the above form with coefficients a_j between 0 and $p-1$ [3, Theorem 1]. This fact means that we can use the p-adic expansion to stand for the whole class of sequences converging to the same p-adic number, just as we use the decimal expansion to stand for the classes of sequences converging to any particular real number.

Definition 6: *The p-adic integers* $\mathbb{Z}_p$. The p-adic integers are defined to be the limits of all Cauchy sequences represented by p-adic expansions of the form

$$a_0 + a_1 \cdot p + a_2 \cdot p^2 + a_3 \cdot p^3 + ... = a_0.a_1a_2a_3\,..._p$$

where the a_i are ordinary integers such that $0 \leq a_i \leq p-1$. We let $\mathbb{Z}_p$ denote the set of all such expressions.

The absolute value of these p-adic integers will be exactly what it was for the regular integers except that now the ord_p will be the power of p dividing a number's p-adic expansion. For example, $|4+3\cdot5+3\cdot5^2+....|_5 = |4.\bar{3}_5|_5 = 1/5^0 = 1$ and $|3\cdot5+3\cdot5^2+....|_5 = 1/5$.

We can add and multiply p-adic integers in a very natural way. For example, we add $4.13\overline{13}_5 + 4.\bar{4}_5$ in the 5-adics as follows:

$$\begin{array}{rrrrrl} & & 1\cdot5 & 1\cdot5^2 & 1\cdot5^3 & 1\cdot5^4 \\ & 4 & +1\cdot5 & +3\cdot5^2 & +1\cdot5^3 & +3\cdot5^4+... \\ + & 4 & +4\cdot5 & +4\cdot5^2 & +4\cdot5^3 & +4\cdot5^4+... \\ \hline & 3 & +1\cdot5 & +3\cdot5^2 & +1\cdot5^3 & +3\cdot5^4+... \end{array}$$

Notice that carrying is done to the right. Similarly, we multiply $4.13\overline{13}_5 \times 4.\bar{4}_5$ in the 5-adics from left to right as follows:

$$\begin{array}{rrrrrl} & 4 & +1\cdot5 & +3\cdot5^2 & +1\cdot5^3 & +3\cdot5^4+... \\ \times & 4 & +4\cdot5 & +4\cdot5^2 & +4\cdot5^3 & +4\cdot5^4+... \\ \hline & 1 & +2\cdot5 & +3\cdot5^2 & +1\cdot5^3 & +3\cdot5^4+... \\ & & 1\cdot5 & +2\cdot5^2 & +3\cdot5^3 & +1\cdot5^4+... \\ & & & 1\cdot5^2 & +2\cdot5^3 & +3\cdot5^4+... \\ & & & & 1\cdot5^3 & +2\cdot5^4+... \\ & & & & & 1\cdot5^4+... \\ \hline & 1 & +3\cdot5 & +1\cdot5^2 & +3\cdot5^3 & +1\cdot5^4+... \end{array}$$

Exercise 7: Find the p-adic expansions of:

1. $2.4\overline{21}_7 + 3.\bar{6}_7$ in $\mathbb{Z}_7$ to four digits,
2. $2.4\overline{21}_7 \times 3.\bar{6}_7$ in $\mathbb{Z}_7$ to four digits.

As we saw above, each positive integer has a p-adic expansion that is just its representation base p. What about negative integers and what about subtraction? If we want to find the 5-adic expansion of -1, we need to find an expression of the form $a_0+a_1\cdot5+a_2\cdot5^2+...$ where $0 \le a_i \le 4$ for all i such that $1+(a_0+a_1\cdot5+a_2\cdot5^2+...) = 0$. Notice that $1+(4+4\cdot5+4\cdot5^2+...) = 0$, hence $4+4\cdot5+4\cdot5^2+...$ is the 5-adic expansion of -1. We can also see that $4.\bar{4}_5 = -1$ by summing a geometric series and using that $\lim_{n\to\infty} 5^n = 0$ as follows:

$$\begin{aligned}4+4\cdot 5+4\cdot 5^2+... &= 4(1+5+5^2+...)\\ &= \lim_{n\to\infty} 4(1+5+5^2+...+5^{n-1})\\ &= \lim_{n\to\infty} 4\left(\frac{1-5^n}{1-5}\right)\\ &= 4\left(\frac{1}{1-5}\right) = -1\end{aligned}$$

Exercise 8: Find the 7-adic expansion of -1.

Question 3: Give a general form for the p-adic expansion of -1 for any prime p.

Exercise 9: Find the 5-adic expansion of -2. You could subtract 2 from $0 = 5+4\cdot 5+4\cdot 5^2+...$ or multiply 2 times the expansion for -1. What is the 7-adic expansion of -2? Find the 5-adic expansion for -405.

Question 4: Given the p-adic expansion of a positive integer, can you write down a formula for the p-adic expansion of its negative?

Exercise 10: Use the geometric series to find the rational numbers with 7-adic expansions $2.4\overline{21}_7$ and $3.\overline{6}_7$. By adding and multiplying these rational numbers, check the calculations in Exercise 7.

However, there are more p-adic integers then there are rational integers, for example, if $p = 5$ then $1/6$ is also a 5-adic integer because it can be written as $1.\overline{4040}_5$. (To check that this expansion represents $1/6$ multiply it by $6 = 1+1\cdot 5$ to get 1.) However, not every rational number is a p-adic integer. The number $36/5 = (1+2\cdot 5+1\cdot 5^2)/5 = 1\cdot 5^{-1}+2+1\cdot 5$ is not a 5-adic integer; it is an element in $\mathbb{Q}_5$ of absolute value 5. It can be shown that $\mathbb{Q}_p$, as we defined it in Definition 5 above, can alternatively be defined as the field of fractions of $\mathbb{Z}_p$.

Now we need to know how to divide p-adic integers. For example, to find the 5-adic expansion of the number $10/75$, we factor out the highest power of p in the numerator and in the denominator so that we need only divide two integers whose p-adic expansions begin with a non-zero first term. In our example, we have

$$\frac{10}{75} = \frac{5 \cdot 2}{5^2 \cdot 3} = 5^{-1}\frac{2}{3}$$

Long division proceeds as usual except that we carry 5's to the right, borrow 5's from the right, and (to make borrowing from the right possible) we write 2 as $2 + 5 \cdot 5 + 4 \cdot 5^2 + 4 \cdot 5^3 + ...$ (remember $5 \cdot 5 + 4 \cdot 5^2 + ... = 0$) as follows:

$$\begin{array}{rrrrrl}
 & 4 & +1 \cdot 5 & +3 \cdot 5^2 & +1 \cdot 5^3 + ... = 4.\overline{13}_5 & \\
\hline
3 \;\big) & 2 & +5 \cdot 5 & +4 \cdot 5^2 & +4 \cdot 5^3 + ... & \\
- & 2 & +2 \cdot 5 & & & \\
\hline
 & & 3 \cdot 5 & +4 \cdot 5^2 & +4 \cdot 5^3 + ... & \\
 & - & 3 \cdot 5 & & & \\
\hline
 & & & 4 \cdot 5^2 & +4 \cdot 5^3 + ... & \\
 & & - & 4 \cdot 5^2 & +1 \cdot 5^3 & \\
\hline
 & & & & 3 \cdot 5^3 + 4 \cdot 5^4 + 4 \cdot 5^5 + ... &
\end{array}$$

Hence, we can think of elements in $\mathbb{Q}_p$ as:

$$\frac{\alpha}{\beta} = \frac{a_0 + a_1 p + a_2 p^2 + ...}{b_0 + b_1 p + b_2 p^2 + ...} = p^{(ord(\alpha) - ord(\beta))}(c_0 + c_1 \cdot p + c_2 \cdot p^2 + ...)$$

where $\alpha \in \mathbb{Z}_p$, $\beta \in \mathbb{Z}_p - \{0\}$ and $c_0 + c_1 \cdot p + c_2 \cdot p^2 + ...$ is an element in $\mathbb{Z}_p$ of absolute value 1. In other words, the first digit in the p-adic expansion of $c_0 + c_1 \cdot p + c_2 \cdot p^2 + ...$ is not 0. From the long division above, we have that $\frac{10}{75} = 5^{-1}(\frac{2}{3}) = 5^{-1}(4.13\overline{13}_5)$. We can of course check this expansion by summing it.

$$4 + 1 \cdot 5 + 3 \cdot 5^2 + 1 \cdot 5^3 + 3 \cdot 5^4 + ... = 4 + 5(1 + 3 \cdot 5)\sum_{i=0}^{\infty} 5^{2i}$$

$$= 4 + 5(16) \lim_{n \to \infty} \frac{1 - 5^{2(n+1)}}{1 - 5^2} = \frac{2}{3}$$

Now we have another way to define the field of p-adic numbers.

Definition 5′: *The field of p-adic numbers,* $\mathbb{Q}_p$ is defined to be the the set of limits of all p-adic Cauchy sequences represented by p-adic expansions of the form

$$p^{-m}(a_0 + a_1 \cdot p + a_2 \cdot p^2 + ...),$$

where the a_i are ordinary integers between 0 and $p-1$.

Exercise 11: Find the p-adic expansions of:

- $4/3$, $3/4$, and $-3/4$ in $\mathbb{Z}_5$,
- $1/6$ in $\mathbb{Z}_7$ and $1/10$ in $\mathbb{Z}_{11}$. Why are these numbers all integers?
- What about $4/30$ and $4/75$ in $\mathbb{Q}_5$? Why are these numbers in $\mathbb{Q}_5$ and not in $\mathbb{Z}_5$?

Question 5: If an element α in $\mathbb{Q}_p$ has p-adic expansion

$$\alpha = p^{-m}(a_0 + a_1 \cdot p + a_2 \cdot p^2 + ...),$$

what is the p-adic expansion of $-\alpha$?

Exercise 12: Find the 7-adic expansion of $\sqrt{2}$ to 3 digits. Note that you need to find a 7-adic number such that

$$(a_0 + a_1 \cdot 7 + a_2 \cdot 7^2 + ...)^2 = 2.$$

There will be two such 7-adic numbers $\sqrt{2}$ and $-\sqrt{2}$. These are not rational numbers so the p-adic digits will not be repeating. Can you find the expansion of $\sqrt{2}$ in $\mathbb{Z}_5$?

Exercise 13: Find the expansion of $\sqrt{-1}$ in $\mathbb{Q}_5$ to three digits. Can you find the expansion of $\sqrt{-1}$ in $\mathbb{Q}_7$?

Question 6: Why do we know that $\sqrt{2}$ will be an integer in $\mathbb{Q}_7$ and $\sqrt{-1}$ will be an integer in $\mathbb{Q}_5$? For which p will $\sqrt{2}$ and/or $\sqrt{-1}$ be in $\mathbb{Q}_p$ and will they always be integers in these fields? Are these p-adic numbers really different from the real numbers and from one another? The rational numbers lie in all these fields.

We visualize the real numbers by picturing them arrayed along a number line. The p-adic integers can be visualized as an infinite system of concentric disks with 0 at the center of all the disks as follows:

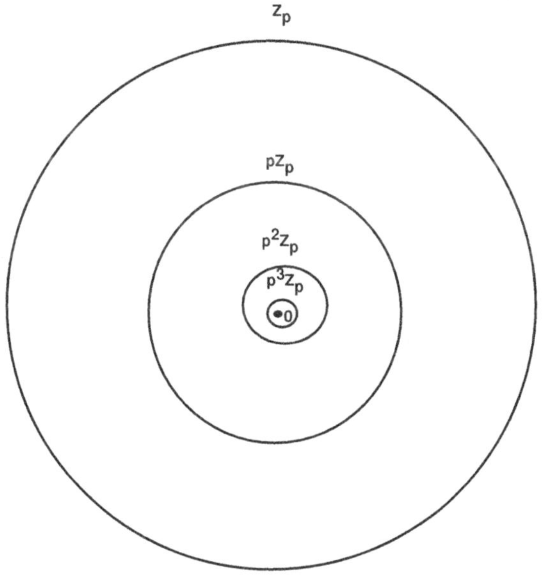

Fig. 8.1: Visualizing $\mathbb{Z}_p$

The elements in $\mathbb{Z}_p$ are guaranteed to have no powers of p in their denominators. The elements inside one disk but not inside the next form a concentric system of annuli. Each annulus contains the elements in $\mathbb{Z}_p$ which are divisible by a fixed number of p's.

An element lies in the outer annulus, that is it lies in $\mathbb{Z}_p$ but not in $p\mathbb{Z}_p$, if it is guaranteed to be divisible by no powers of p. The a_0 in the p-adic expansion $(a_0 + a_1 \cdot p + a_2 \cdot p^2 + ... + a_n \cdot p^n + ...)$ for such an element can be any number between 1 and $p - 1$ but cannot be 0. The elements in this annulus all have absolute value equal to 1 since they are not divisible by any powers of p.

An element lies inside the second annulus, that is inside $p\mathbb{Z}_p$ but not inside $p^2\mathbb{Z}_p$, if it is guaranteed to be divisible by p but not divisible by p^2. For such an element, the a_0 in its p-adic expansion equals 0 and the a_1 cannot be 0. The elements in this annulus all have absolute value equal to p^{-1} since they are all divisible by exactly one power of p.

Therefore as we work our way into the center of the $\mathbb{Z}_p$ disk, all the elements in the same annulus have the same absolute value and the absolute values are decreasing powers of p. If an element is in the disk $p^n\mathbb{Z}_p$ but not in $p^{n+1}\mathbb{Z}_p$ then its absolute value is p^{-n}. The smallest element is 0 at the center of all the disks with absolute value equal to 0. The visualization of $\mathbb{Z}_p$ (with absolute value thought of as height above 0) now looks something like an infinite upside down wedding cake with 0 way down at the center. The levels representing the widening annuli increase in height by powers of p until the outermost annulus in $\mathbb{Z}_p$ which is at height 1.

We use the notation $p\mathbb{Z}_p$ for the elements in the first disk since they all look like p times a general element in the p-adic integers and have expansions of the following form:

$$p(a_0 + a_1 \cdot p + ... + a_n \cdot p^n + ...).$$

An element lies in the second disk if it is divisible by p^2. Hence, we use $p^2\mathbb{Z}_p$ to denote elements in that disk. Of course, all the elements which are in the second disk are also included in the first disk. This inclusion explains the concentric nature of the disks.

If $p = 5$, there are four types of elements in $\mathbb{Z}_5$ but not in $5\mathbb{Z}_5$ and they are:

- $1 + a_1 \cdot 5 + a_2 \cdot 5^2 + ..$
- $2 + a_1 \cdot 5 + a_2 \cdot 5^2 + ..$
- $3 + a_1 \cdot 5 + a_2 \cdot 5^2 + ..$
- $4 + a_1 \cdot 5 + a_2 \cdot 5^2 + ...$

We can denote these types of elements as $1 + 5\mathbb{Z}_5$, $2 + 5\mathbb{Z}_5$, $3 + 5\mathbb{Z}_5$, and $4 + 5\mathbb{Z}_5$ and think of them as smaller disks of the same size as $5\mathbb{Z}_5$ centered around their first coefficient and lying in the outer annulus. There are no other elements in the annulus $\mathbb{Z}_5 - 5\mathbb{Z}_5$, so the space between the disks in the annulus is empty. Of course, $5\mathbb{Z}_5$ is just $0 + 5\mathbb{Z}_5$. All of these 5 disks have the same number of elements in them and so should be of the same size or diameter.

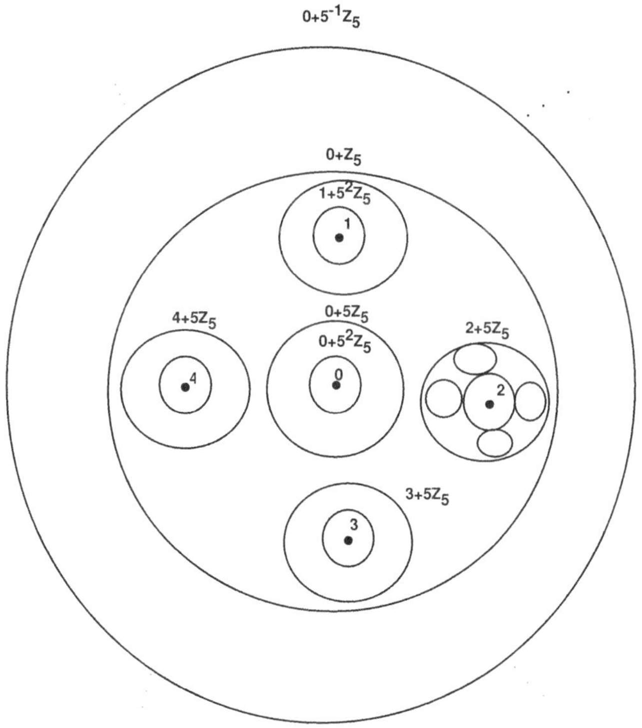

Fig. 8.2: Visualizing $\mathbb{Z}_5$

If we consider the disk $1 + 5\mathbb{Z}_5$, it is composed of an infinite system of concentric disks with 1 at its center. We will use the notation $1 + 5^n\mathbb{Z}_5$ where $n \geq 1$ to denote these concentric disks. Just as the disks about 0 were contained in one another, we have that $1 + 5\mathbb{Z}_5$ contains $1 + 5^2\mathbb{Z}_5$ which contains $1 + 5^3\mathbb{Z}_5$, and so on. In the outer annulus about 1, we have the elements in $1 + 5\mathbb{Z}_5$ but not in $1 + 0 \cdot 5 + 5^2\mathbb{Z}_5$. This annulus is completely covered by four smaller disks of the same size as $1 + 5^2\mathbb{Z}_5$. These four disks contain the elements of the form

- $1 + 1 \cdot 5 + 5^2\mathbb{Z}_5$

- $1 + 2 \cdot 5 + 5^2\mathbb{Z}_5$
- $1 + 3 \cdot 5 + 5^2\mathbb{Z}_5$
- $1 + 4 \cdot 5 + 5^2\mathbb{Z}_5$.

Each of these disks is composed of more concentric disks centered about $1 + a_1 \cdot 5$ of the form $1 + a_1 \cdot 5 + 5^n\mathbb{Z}_5$ where $n \geq 2$. These disks get smaller and smaller in size and they are the analogues to the open intervals on the real number line.

Question 7: When describing the model for $\mathbb{Z}_p$, we thought of the disks of the form $\alpha + p^n\mathbb{Z}_p$ as getting smaller and smaller as n gets bigger. Suppose we have two elements in $1 + 5\mathbb{Z}_5$, say $1 + 4 \cdot 5 + 4 \cdot 5^2 + 4 \cdot 5^3 + ...$ and $1 + 3 \cdot 5 + 3 \cdot 5^2 + 3 \cdot 5^3 + ...$, what is the distance between them? What is the distance between $1 + 4 \cdot 5 + 4 \cdot 5^2 + 4 \cdot 5^3 + ...$ and $1 + 4 \cdot 5 + 3 \cdot 5^2 + 3 \cdot 5^3 + ...$? What is the maximal distance between any two points in the disk $1 + 5\mathbb{Z}_5$? This maximal distance is the diameter of the disk. What is the diameter of a disk centered at α of the form $\alpha + p^n\mathbb{Z}_p$ where α is any fixed element in $\mathbb{Z}_p$ and n is a fixed positive integer?

Question 8: Complete the above picture of $\mathbb{Z}_5$ by filling in all the remaining disks of the same size as $5^2\mathbb{Z}_5$. Make a similar diagram for $\mathbb{Z}_7$. Describe the diagram in general for $\mathbb{Z}_p$.

Question 9: Notice that the set of all points of distance less than or equal to 1 away from 0, $\{x \in \mathbb{Q}_p : |x - 0|_p \leq 1\}$, is precisely all of $\mathbb{Z}_p$. Hence, we say that $\mathbb{Z}_p$ is the ball of radius 1 centered at 0. What is the set of all points of distance less than or equal to 1 away from 1, $\{x \in \mathbb{Q}_p : |x - 1|_p \leq 1\}$. This set must be the the ball of radius 1 centered at 1. What about the ball of radius 1 around 2? How can a ball have more than one center? Is the radius of a p-adic ball the same as its diameter? Show using the properties of the p-adic absolute value that any point in a ball is its center.

Glancing back at Figure 8.2, we should not think of 0 as the "center" of $\mathbb{Z}_5$ any longer (although it is convenient to put it in the center because of the value of the absolute value in the annuli). Instead we should imagine the five balls of the form $a + 5\mathbb{Z}_5$ floating in the $\mathbb{Z}_5$ circle so that any one of them can be pushed into the center.

Question 10: In the picture of $\mathbb{Z}_5$ above, there is a disk outside and containing $\mathbb{Z}_5$ denoted by $5^{-1}\mathbb{Z}_5$. Describe the elements inside that disk but not inside $\mathbb{Z}_5$. What would their absolute values be?

Outside and containing the disk $5^{-1}\mathbb{Z}_5$, there is a disk which should be denoted by $5^{-2}\mathbb{Z}_5$. In this way, we see another ever-widening system of annuli working their way out from $\mathbb{Z}_5$. The union of all the elements in these disks is the entire field of 5-adic numbers, $\mathbb{Q}_5$.

8.5 References

George Bachman. *Introduction to p-Adic Numbers and Valuation Theory.* Academic Press, 1964.

Fernando Q. Gouvêa. *p-adic Numbers.* Springer-Verlag, 1993.

Kelsey Houston-Edwards. *An Infinite Universe of Number Systems*, Quanta Magazine, Oct. 19, 2020.
https://www.quantamagazine.org/peter-scholze-and-the-future-of-arithmetic-geometry-20160628/

Erica Klarreich. *The Oracle of Arithmetic*, Quanta Magazine, June 28, 2016.
https://www.quantamagazine.org/how-the-towering-p-adic-numbers-work-20201019/

Neal Koblitz. *p-adic Numbers, p-adic Analysis, and Zeta-Functions.* Springer-Verlag, 1977.

Neal Koblitz. *The p-adic approach to solutions of equations over finite fields*, Amer. Math. Monthly **87** (1980) no. 2, 115-118.

Kurt Mahler. *p-adic Numbers and their Functions.* Cambridge University Press, 1973.

B. Mazur. *On the passage from local to global in number theory*, Bull. Amer. Math. Soc. **29** no. 1, 14-50.

8.6 Computer programs

8.6.1 *Mathematica* programs

Factoring Integers

The following command displays the prime factorization of an integer.

```
PrimeFactorization[nr_]:=Module[{fd}, fd = FactorInteger[nr];
Print[nr, " = ", If[Length[fd] > 1,
    CenterDot@@(Superscript@@@fd), (Superscript@@fd[[1]])]]]
```

```
PrimeFactorization[99999]
```

$99999 = 3^2 \cdot 41^1 \cdot 271^1$

The command also works for rational numbers:

```
PrimeFactorization[1000/555]
```

$\frac{200}{111} = 2^3 \cdot 3^{-1} \cdot 5^2 \cdot 37^{-1}$

Computing p-adic Norms

The next command computes the p-adic norm of a rational number x.

```
PadicNorm[x_, p_ : 5]:=Module[{a = Cases[FactorInteger[x], {p, _}]},
    If[a==={}, If[x == 0, 0, 1], First[1/Power@@@a]]]
```

PadicNorm[1/15, 3]

3

PadicNorm[18, 3]

$\frac{1}{9}$

PadicNorm[21/20, 3]

$\frac{1}{3}$

PadicNorm[2, 7]

1

The default for p is 5.

PadicNorm[21/20]

5

The last command is adapted from a notebook by Eric W. Weisstein, see: http://mathworld.wolfram.com/notebooks/Algebra/p-adicNorm.nb

Checking Divisibility

The command below checks for which integer values of x, theexpression $3x-2$ is divisible by 5. (You can change the "3x-2" and the "5").

```
Solve[3x - 2 == 0, Modulus -> 5][[1]]
```

$\{x \to 4\}$

Chapter 9

Parametric Curve Representation

9.1 Introduction

Look at Figure 9.1 below. Its shape is pleasing partly because it is so nicely

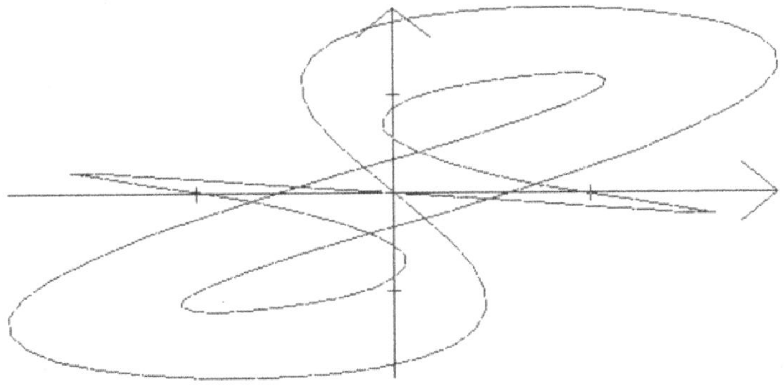

Fig. 9.1: A sample curve

symmetric. In this chapter you will learn how to create pictures like this one and determine—even predict—their symmetries. The ideas and language of symmetry permeate mathematics, physics and chemistry. The pictures you will draw are *curves* described by *parametric functions.* Curves and their descriptions occur in many areas of mathematics and science. Such curves, for example, can describe the motion of a particle or the trajectory of a spaceship. Computer graphics uses parametric representations of curves in fundamental ways. For example, they are used in drawing circles and ellipses, or in creating "wire-frame" pictures of solid surfaces. You also see such curves and surfaces in two and three dimensions in the study of multivariable calculus.

Looking for patterns among the examples will be one of your goals in this chapter. Carefully organizing your experimental findings will help you identify patterns and the conditions that cause the patterns to occur.

9.2 Symmetries and closed curves

9.2.1 Definitions

Imagine that a bug crawls along an x-y plane and that we are to trace the path that the bug takes. At any time t, the bug is at a point (x, y), depending on t. As t changes, the coordinates of the bug's position can be represented as $(x(t), y(t))$, where $x(t)$ and $y(t)$ are functions. If the bug doesn't hop around, we can even believe that the functions are continuous. So we will make our formal definition with this in mind.

A curve in the x-y plane can often be represented as a pair of continuous functions

$$x = x(t), \quad y = y(t)$$

over an interval $a \leq t \leq b$. As the value of t varies from a to b, the point (x, y) traces out a curve in the plane. Technically, the *continuous curve*, or just *curve* corresponding to the continuous functions x and y is the set of ordered pairs

$$\Phi = \{(x(t), y(t)) : a \leq t \leq b\}.$$

The set of points in the plane forming the curve is called its *graph.* The pair of functions $x(t)$ and $y(t)$ are called *parametric functions* for the curve Φ, the variable t is called a *parameter*, and we say that a curve described in this way is *represented parametrically* by the two functions. Similar definitions can describe curves parametrically in three or more dimensions.

We need language to describe the geometric features of the curve Φ.

- We call $(x(a), y(a))$ the *initial point* and $(x(b), y(b))$ the *terminal point* of Φ.

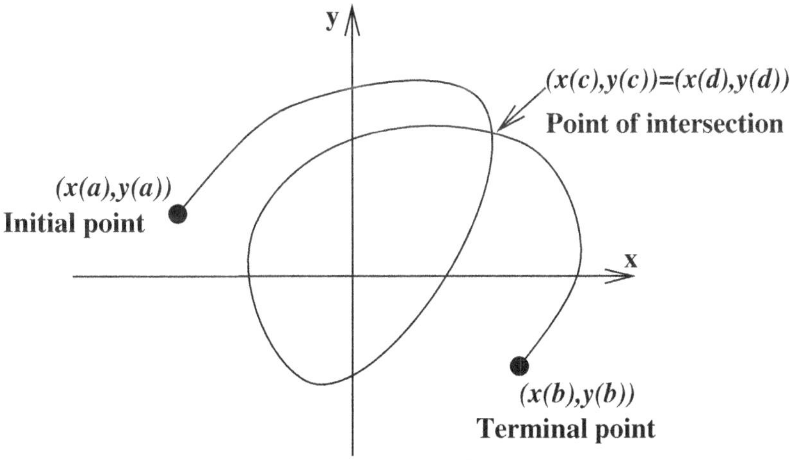

Fig. 9.2: Initial, terminal and intersection points

- We say that the curve Φ *intersects itself* at c and d ($c \neq d$) provided $x(c) = x(d)$ and $y(c) = y(d)$.
- The curve Φ is *closed* if $(x(a), y(a)) = (x(b), y(b))$, and it is a *simple* closed curve if it is closed and does not intersect itself (except at the initial and terminal points).
- The curve Φ, defined on the interval $a \leq t \leq b$, is called *finitely intersecting* (or *fi*) provided there are only a finite number of places where Φ intersects itself.

Can you think of an example of a curve which is not finitely intersecting?

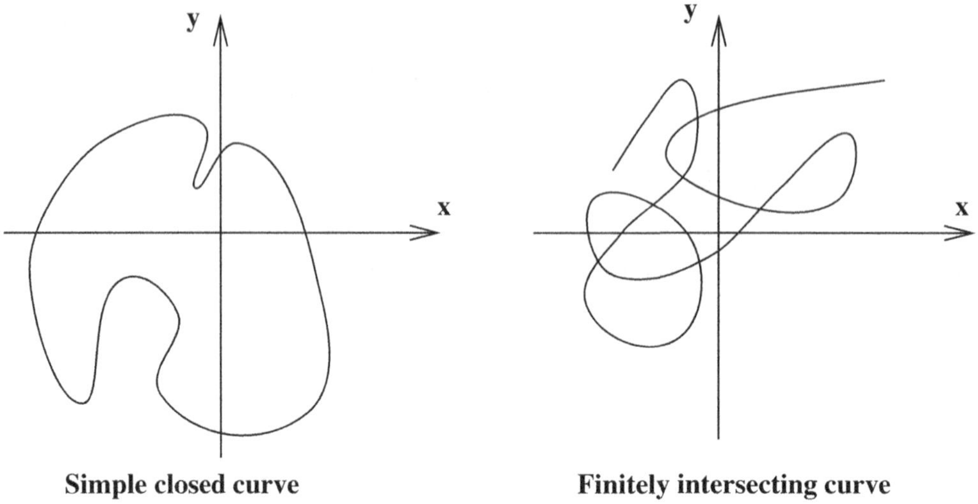

Fig. 9.3: Two curve examples

Exercise 1: Look at the curve

$$x(t) = \cos(t)$$
$$y(t) = \sin(2t)$$

Using a calculator, make a table of values of x and y against t, for $t = 0, 0.2, 0.4, 0.6, ..., 3.1, 3.2$. On a piece of graph paper, plot the curve by first plotting the points, (x, y), and then connecting the dots. From your table of values, you should see reasonable bounds on x and y to use for the axes. Is this curve closed? Is it a simple closed curve? Is it finitely intersecting?

9.2.2 Graphing with the computer

Plotting parametric functions by hand is very tedious, and the computer or a graphing calculator can be extremely helpful. Free online web apps like *Desmos* and *Geogebra* readily plot parametric curves and so does *Mathematica* (see code at end of chapter). For example, to plot our first example below parametrically using *Desmos'* graphing calculator, which can be accessed online at https://www.desmos.com/calculator, enter the formula:

```
(cos(t),sin(t))
```

Be sure to enter the range of parameter $0 \leq t \leq 2\pi$. When you type *2pi*, *Desmos* will automatically convert it to 2π.

An important note here: Many plotting programs simply show you the graph of the parametric functions rather than showing *how* the curve is traced out as t varies. To see how the curve is traced out in *Desmos*, change the parameter range to $0 \leq t \leq T$. This will prompt you to create a slider for T. As you adjust the slider you will see the curve traced out. Set the range of the slider by clicking on the $T =$ to be sure that you see the entire graph. You can also set the step size to a small number in order to get a slower animation which will allow you to see what is happening. Many parametric curves trace back over themselves, so it is useful to also see the point on the graph corresponding to T. To do this in *Desmos* enter on another line:

```
(cos(T),sin(T))
```

You can get an animation by pressing the play button to the left of the slider. This example is shown in Figure 9.4.

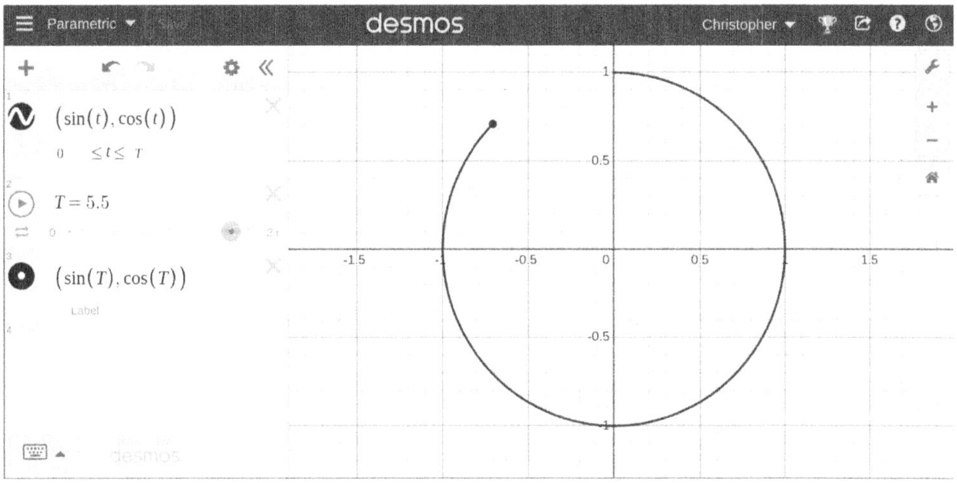

Fig. 9.4: Desmos tracing out the unit circle using a parametric plot

Use a graphing program to plot the following curves. You will need to change the functions $x = x(t)$ and $y = y(t)$ as needed, and you should be sure that the interval $[a, b]$ is properly specified. In each case, describe the curve that the equations generate. In the first two examples, describe which properties of the functions produce the curves that you get. Can you do the same for the third example?

Example 1:

$$x(t) = \cos(t)$$
$$y(t) = \sin(t) \quad \text{over} \quad 0 \le t \le 2\pi.$$

Example 2:

$$x(t) = \ t$$
$$y(t) = \sin(t) \quad \text{over} \quad 0 \le t \le 2\pi.$$

Example 3:

$$x(t) = \sin(t) + cos(t)$$
$$y(t) = \sin(4t) + cos(2t) \quad \text{over} \quad 0 \le t \le 2\pi.$$

Exercise 2: What symmetries do you see in these three examples? What does *symmetry* mean in this setting? Describe the symmetries you see as precisely as you can. Think carefully about these questions before moving on to the next section.

9.2.3 Symmetry in curves

Curves can be symmetric in various ways.

- We say two points P and Q are *symmetric about a line* ℓ if ℓ is the perpendicular bisector of the line segment between P and Q, and in this case we say that Q *is symmetric to* P *across* ℓ. Then a curve Φ is *symmetric about a line* ℓ if whenever a point P is on Φ, so is the point Q symmetric to P across ℓ. The line ℓ is called a *mirror line* for Φ.
- Similarly, we say two points P and Q are *symmetric about a point* M if M bisects the line segment between P and Q; we also say that Q *is symmetric to* P *through* M. Then a curve Φ is *symmetric about a point* M if whenever a point P is on Φ, then so is the point Q that is symmetric to P through M. In this case, the point M is called a *center of symmetry* for Φ.

We will now focus on a special situation—namely when the functions to be plotted are of the form

$$\begin{aligned} x(t) &= \sin(pt) + \cos(qt) \\ y(t) &= \sin(rt) + \cos(st) \end{aligned} \tag{9.1}$$

where $0 \leq t \leq 2\pi$, and p, q, r, s are all positive integers. In this lab, we shall denote this system by the *4-tuple* (p, q, r, s). Later we'll modify our requirement that p, q, r, and s be non-zero.

Exercise 3: Describe the curves produced by the following parametric functions: (A), (B), and (C). In each case, pay attention both to how the curve is traced out and to what its final appearance is. Do you see any relationships among these graphs? The curves are all defined over the interval $0 \leq t \leq 2\pi$.

$$\text{A.} \quad \begin{aligned} x &= \sin(t) + \cos(2t) \\ y &= \sin(t) + \cos(t) \end{aligned}$$

$$\text{B.} \quad \begin{aligned} x &= \sin(2t) + \cos(4t) \\ y &= \sin(2t) + \cos(2t) \end{aligned}$$

$$\text{C.} \quad \begin{aligned} x &= \sin(3t) + \cos(2t) \\ y &= \sin(5t) + \cos(4t) \end{aligned}$$

Of course, these examples could be denoted, respectively, as $(1, 2, 1, 1)$, $(2, 4, 2, 2)$, and $(3, 2, 5, 4)$, as described under equations (10.1).

If we start with a set G of points in the plane, we can ask whether there is an fi-closed curve Φ whose graph is the set G. This question gives rise to the following definition.

- We say that a set G of points in the plane *depicts an fi-closed curve* if there is a finitely intersecting closed curve Φ whose graph is G.

This definition is very specific, and it is possible for a curve to be closed while its graph does not depict an fi-closed curve. What is an example that illustrates this?

We can generalize the concept of a parametric representation of a curve in the plane to that of a curve in k-dimensional space by using k continuous functions rather than just 2. For example, if $k = 3$ we obtain a curve in space that might describe, for instance, the trajectory of a space ship on a mission to the moon.

9.3 Questions to explore

One way to get a closed curve is to choose functions x and y which are periodic with a common period. Then the graph over an interval corresponding to the common period will surely be closed.

Exercise 4: Why is this true?

Question 1: Plot the function $(p, q, r, s) = (2, 6, 2, 2)$; that is, plot

$$\begin{aligned} x(t) &= \sin(2t) + \cos(6t) \\ y(t) &= \sin(2t) + \cos(2t) \quad \text{over} \ \ 0 \leq t \leq 2\pi. \end{aligned}$$

Describe this curve in your own words, and answer the following questions about it:

(a) Is this an fi-closed curve?

(b) Does this graph depict an fi-closed curve?

(c) Is it symmetric about the x-axis?

(d) Is it symmetric about the y-axis?

(e) Is it symmetric about the origin?

(f) Is it symmetric about any other point or line?

(g) Do you notice other characteristics of this curve?

Solution: Here is a description of what happens in this example.

This is a closed curve which is traversed twice (over $[0, 2\pi]$). Hence, as a curve, it intersects itself infinitely many times: for instance, at any values w and $w + \pi$, for any value w in $[0, \pi]$. As a graph, however, it depicts an fi-closed curve that intersects itself five times. What are the functions $x(t)$ and $y(t)$ that would form that fi-closed curve with the same graph?

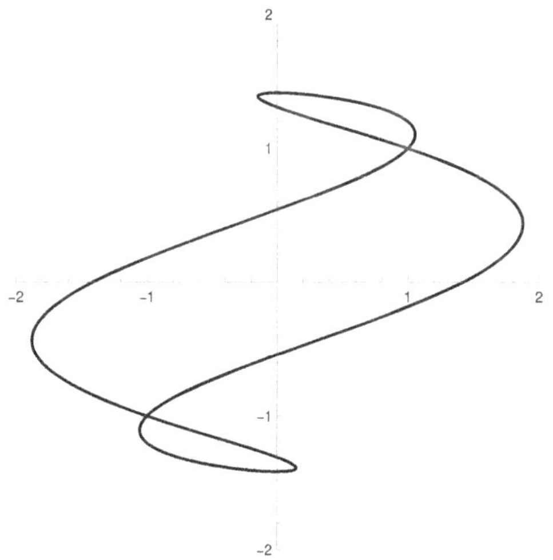

Fig. 9.5: The closed curve (2,6,2,2)

You might further organize your observations in a table like the one that follows.

(p, q, r, s)	fi-closed curve	Depicts fi-closed curve	mirror lines			Centers of symmetry	
			x-axis	y-axis	other	origin	other
(2,6,2,2)	NO	YES	NO	NO	NO	YES	NO

Question 2: Repeat Question 1 for the following parametrically-described curves.

A. $x = \sin(4t) + \cos(t)$, $y = \sin(t) + \cos(2t)$　B. $x = \sin(4t) + \cos(t)$, $y = \sin(t) + \cos(4t)$

C. $x = \sin(t) + \cos(t)$, $y = \sin(t) + \cos(t)$　D. $x = \sin(t) + \cos(t)$, $y = \sin(t) + \cos(5t)$

E. $x = \sin(5t) + \cos(2t)$, $y = \sin(3t) + \cos(2t)$　F. $x = \sin(7t) + \cos(8t)$, $y = \sin(3t) + \cos(2t)$

G. $x = \sin(6t) + \cos(3t)$, $y = \sin(4t) + \cos(t)$　H. $x = \sin(12t) + \cos(9t)$, $y = \sin(3t) + \cos(6t)$

I. $x = \sin(5t) + \cos(7t)$, $y = \sin(2t) + \cos(6t)$　J. $x = \sin(6t) + \cos(8t)$, $y = \sin(5t) + \cos(t)$

For the curves (A) - (J), make a chart with the same headings we used in Question 1:

	(p, q, r, s)	fi-closed curve	Depicts fi-closed curve	mirror lines x-axis	mirror lines y-axis	mirror lines other	Centers of symmetry origin	Centers of symmetry other
A	(4,1,1,2)							
B	(4,1,1,4)							
C								

Look over your observations for the examples (A) – (J). Can you formulate a conjecture as to what properties of these parametric functions enable you to say yes to questions (a) - (f)? That is, what patterns do you see in the integers p, q, r and s that yield these results? You can (and should) try lots of other integers here to help yourself see regularities and refine your conjectures. Try to prove that your conjectures are, in fact, true.

9.4 Polar representation of curves

Example: Look at the parametric curves (over the interval, $0 \leq t \leq 2\pi$):

A. $x(t) = (1 + 2\cos(t))\cos(t)$
$y(t) = (1 + 2\cos(t))\sin(t)$

B. $x(t) = (1 - 2\sin(t))\cos(t)$
$y(t) = (1 - 2\sin(t))\sin(t)$

C. $x(t) = \cos(t)^2$
$y(t) = \cos(t)\sin(t)$

Use your parametric function plotter to plot these curves and describe them in terms of the properties you looked at in the previous section (using questions (a) - (g)).

These parametric equations are examples of functions represented in *polar coordinates.* That is

$$\begin{aligned} x(t) &= r(t)\cos(t) \\ y(t) &= r(t)\sin(t) \end{aligned} \tag{9.2}$$

where $r = r(t)$ is a function of the real variable t.

Question 3: Suppose that r and t are real numbers. Let $x = r\cos(t)$ and $y = r\sin(t)$.

1. What is the distance from the point (x, y) to the origin?
2. What is the *tangent* of the angle that the line from the origin to (x, y) makes with the positive x-axis? Hence, what *is* that angle?

The numbers r and t are called the *polar coordinates* of the point (x, y). (The numbers x and y are the *cartesian coordinates* of the point.) A function $r = r(t)$ represented in this way (see equations (9.2)) is said to be in *polar coordinates.*

You can, of course, graph these functions parametrically as done above. Or, you can plot them directly as polar equations. For example, to plot example (A) in *Desmos* enter:

```
r(theta)=1+2cos(theta)
```

Notice that *Desmos* automatically changes the word *theta* into the Greek letter θ. Verify that the results are the same as those obtained using a parametric function plotter. Now try examples (B) and (C) in polar coordinates and compare to your earlier graphs.

Do some more experimentation by trying these additional polar curves.

D. $r = 1 + 3\cos(2t) + 2\sin(3t)$ "Moose head curve"
E. $r = 2 + 3\cos(2t) + 2\sin(3t)$
F. $r = \cos(t)$
G. $r = \cos(2t)$
H. $r = \cos(3t)$
I. $r = \cos(4t)$
J. $r = \cos(5t)$

Question 4: What pattern seems to emerge in the geometry of examples (F) – (J)? Can you prove it? Find some more interesting curves in polar coordinates. Write down their formulas and describe their main features.

9.4.1 Parameterized families of polar curves

We sometimes come upon families of functions that change as certain conditions which define them change. For instance, the polar function $r(t) = 1 - 2\sin(t)$ is one member of a *parameterized family* of curves

$$r(t) = 1 - c\sin(t)$$

where c can be any real number. Be careful not to confuse the term "parameterized family" with "parametric equations" or "parametric representations" for a single curve. For each fixed value of c, as we vary the parameter t, we trace out a *single* curve. That describes the curves we have been studying in the earlier sections. In this case, we have a family of *many* curves, one for each value of the "parameter" c.

Question 5: Using a polar function plotter, describe the graphs of the various functions you get for $r(t) = 1 - c\sin(t)$ by letting c vary over the interval $-1 \le c \le 1$. For instance, let $c = 1$, $c = 0.7$, $c = 0.5$, $c = 0.2$, $c = 0$, $c = -0.2$, ... , $c = -1$. Sketch each of these graphs and observe how each one is related to the previous one and to the group as a whole. What symmetries do you notice? What effect does increasing/decreasing the parameter c have on the shape and size of the polar curve that the function traces?

Question 6: Repeat the previous exploration for values of c which are greater than 1; try $c = 2$, $c = 5$, $c = 20$, $c = 200$. What do you observe? Can you determine why this behavior occurs? What aspects of the function become dominant and force this behavior as c increases? As c decreases?

9.5 Additional ideas to explore

There are many directions that the computer will allow you to explore using the kinds of functions we have introduced here. For example:

1. If a parametrically expressed curve depicts an fi-closed curve yet is *not* an fi-closed curve, find a parametrically defined function that is fi-closed and has the same graph.

2. Why did some of the functions you studied draw a curve and then proceed to trace over the same curve again and again? Under what conditions does it draw a curve just once?

3. A parametric function of the form (p, q, q, p) (where p and q are *positive* integers) represents a "rose" graph of several "petals." Under what condition does this trace the rose graph just once? If this condition is satisfied, can you predict how many petals the rose will have? Can you prove this?

4. Experiment with curves on intervals other than $0 \leq t \leq 2\pi$.

5. Experiment with curves (p, q, r, s) where some of these integers are zero. Which of the properties we have looked at still hold or do not hold? You can look, for example, at "parity-like" questions where the mappings are represented by *4-tuples* of d's, e's, and 0's (with d standing for "odd" and e for "even").

6. What happens if some of p, q, r, or s are not integers?

7. Experiment with parametrically defined functions that are completely different from those we have looked at in this chapter.

9.6 Computer programs

9.6.1 *Mathematica* programs

Static Plots

General Plot

The following shows how to draw general parametric plots, with comments briefly explaining each piece of the command.

```
ParametricPlot[
  {t, Sin[t]}, (*x − function, y − function*)
  {t, 0, 2 * Pi}, (*parameter and its range*)
  PlotRange → {{−2, 2}, {−2, 2}}, (*limits of graphing window*)
  AspectRatio → 1, PlotPoints → 1000, PlotStyle → Directive[Thick, Black],
  AxesLabel → {x, y}](*technical stuff*)
```

Here's the same command, without the comments, to make it easier for you to modify and draw your own parametric plots.

```
ParametricPlot[{Cos[t], Sin[t]}, {t, 0, 2 * Pi}, PlotRange → {{−2, 2}, {−2, 2}},
    AspectRatio → 1, PlotPoints → 1000, PlotStyle → Directive[Thick, Black],
    AxesLabel → {x, y}]
```

Shortcut: Param2

Param2, below, will make it easy for us to investigate the curves described in Section 9.2.3, namely, the parametric curves $x = \sin(pt)\cos(qt)$, $y = \sin(rt)\cos(st)$ over the interval $0 \leq t \leq 2\pi$. Note that you must define it before running it!

```
Param2[p_, q_, r_, s_]:=ParametricPlot[{Sin[p * t] + Cos[q * t],
    Sin[r * t] + Cos[s * t]}, {t, 0, 2 * Pi}, PlotRange → {{−2, 2}, {−2, 2}},
    AspectRatio → 1, PlotPoints → 1000, PlotStyle → Directive[Thick, Black],
    AxesLabel → {x, y}];

Param2[6, 8, 5, 1]
```

Animation

The following command will animate a parametric plot. You can modify: the x-function; the y-function; the interval for the parameter; and the range of the graphing window.

```
Clear[x, y, t, starttime, endtime, win];
x[t_] = Cos[3t] + Sin[6t];
y[t_] = Sin[4t] + Cos[t];
starttime = 0;
endtime = 2 * Pi;
```

```
win = 2;

(*Do not change anything below this line*)
Animate[Show[
    ParametricPlot[{x[u], y[u]}, {u, starttime, endtime},
    PlotRange → {{−win, win}, {−win, win}},
    PerformanceGoal → "Quality", AxesLabel → {x, y}],
    ParametricPlot[{x[u], y[u]}, {u, starttime, t},
    PlotRange → {{−win, win}, {−win, win}},
    PerformanceGoal → "Quality", PlotStyle → Directive[Thick, Black]],
    Graphics[{Red, Thickness[.01], Circle[{x[t], y[t]}, .02]}],
    Graphics[Text[StringJoin["t = ",
    ToString[NumberForm[t/Pi, {2, 1}]], ToString[Pi, StandardForm]],
    {.75win, .95win}]]], {t, starttime + (endtime − starttime)/10.^5, endtime},
    AnimationRepetitions → 1, AppearanceElements → "ResetButton",
    ControlPlacement → Bottom]
```

To change the size of the graphing window, click on the graph (until you get a yellow box), grab a corner, and resize it.

Here is a shorter version for animations - you can only change p, q, r and s to graph $x = \sin(pt)\cos(qt)$, $y = \sin(rt)\cos(st)$ over the interval $0 \leq t \leq 2\pi$:

```
Anima2[p_, q_, r_, s_]:=Module[{}, {x[t_], y[t_]} = {Sin[pt] + Cos[qt], Sin[rt] + Cos[st]};
Animate[Show[
    ParametricPlot[{x[u], y[u]}, {u, 0, 2π}, PerformanceGoal → "Quality",
    AxesLabel → {x, y}],
```

```
ParametricPlot[{x[u], y[u]}, {u, 0, t}, PerformanceGoal → "Quality",
PlotStyle → Directive[Thick, Black]],
Graphics[{Red, Thickness[.01], Circle[{x[t], y[t]}, .02]}],
Graphics[Text[StringJoin["t = ", ToString[NumberForm[t/Pi, {2, 1}]],
ToString[Pi, StandardForm]], {.75win, .95win}]]], {t, 2π/10.^5, 2π},
AnimationRepetitions → 1, AppearanceElements → "ResetButton",
ControlPlacement → Bottom]]
```

```
Anima2[6, 8, 5, 1]
```

Polar Coordinates

The following shows how to draw general polar coordinate plots, with comments briefly explaining each piece of the command.

```
r[t_] = Cos[3t]; (*the radius r(t) as a function of t*)
ParametricPlot[{r[t]Cos[t], r[t]Sin[t]},
    {t, 0, 2 * Pi}, (*parameter and its range*)
    PlotRange → {{−3, 3}, {−3, 3}}, (*limits of graphing window*)
    AspectRatio → 1, PlotPoints → 1000, PlotStyle → Directive[Thick, Black],
    AxesLabel → {x, y}]
    (*technical stuff*)
```

Here's the same command, without the comments, to make it easier for you to modify and draw your own polar coordinate plots:

```
r[t_] = 1 + 2Cos[t];
ParametricPlot[{r[t]Cos[t], r[t]Sin[t]}, {t, 0, 2 * Pi}, PlotRange → {{ 3, 3}, {−3, 3}},
    AspectRatio → 1, PlotPoints → 1000, PlotStyle → Directive[Thick, Black],
    AxesLabel → {x, y}]
```

Here is the corresponding animation:

```
Clear[r, x, y, t, starttime, endtime, win];
r[t_] = 1 + 2Cos[t];
starttime = 0;
endtime = 2 * Pi;
win = 4;

(*Do not change anything below this line*)
x[t_] = r[t]Cos[t]; y[t_] = r[t]Sin[t];
Animate[Show[
    ParametricPlot[{x[u], y[u]}, {u, starttime, endtime},
    PlotRange → {{−win, win}, {−win, win}},
    AxesLabel → {x, y}, PerformanceGoal → "Quality"],
    ParametricPlot[{x[u], y[u]}, {u, starttime, t},
    PlotRange → {{−win, win}, {−win, win}},
    PerformanceGoal → "Quality", PlotStyle → Directive[Thick, Black]],
    Graphics[{Dotted, Gray, Line[{{0, 0}, {x[t], y[t]}}]}],
    Graphics[{Red, Thickness[.01], Circle[{x[t], y[t]}, .02]}],
    Graphics[Text[StringJoin["t = ", ToString[NumberForm[t/Pi, {2, 1}]],
    ToString[Pi, StandardForm]], {.75win, .95win}]]],
    {t, starttime + (endtime − starttime)/10.^5, endtime}, AnimationRepetitions → 1,
    AppearanceElements → "ResetButton", ControlPlacement → Bottom]
```

Chapter 10

Numerical Integration

10.1 Introduction

Many real phenomena are described in terms of differential equations, so many questions of great interest involve integration. In this chapter you will examine several numerical approaches to calculating integrals, some perhaps familiar, and others new. The emphasis will be on examining the accuracy and efficiency of different methods.

Recall that if a and b are definite numbers and $f(x)$ is a "nice" function on $[a, b]$, then $\int_a^b f(x)\,dx$ is a number, called the *integral* of $f(x)$ over the interval $[a, b]$. This number can be computed using various methods:

1. Directly from the definition, by taking a limit of approximating sums.
2. From the fundamental theorem of calculus, by seeking a function $F(x)$ whose derivative is $f(x)$; the required number is then $F(b) - F(a)$.
3. Using a method for solving differential equations to compute $y(b)$ where $y(t)$ is the solution of the initial value problem $y(a) = 0$, $dy/dt = f(t)$.
4. Hit-or-miss methods using cleverness and more advanced mathematics (especially power series and complex variables).

If it is possible to evaluate an integral by hand, without a computer or calculator, then it is usually done by method 2 or method 4. Since method 4 is anything but systematic, and there is no guarantee of success, most calculus courses spend a lot of time on method 2 and systematic ways to find antiderivatives. Unfortunately, *most* functions do not have antiderivatives given by nice formulas. To be more precise, most functions which can be expressed as sums, products, quotients, powers and compositions of the functions

$$f(x) = \text{constant}, \ x, \ \sin x, \ \cos x, \ e^x, \ \ln(x),$$

$$\sin^{-1}(x), \ \cos^{-1}(x), \ \tan^{-1}(x), \ ...$$

do not have antiderivatives that can be so expressed. Explicit examples are

$$f(x) = e^{-x^2}, \ f(x) = \frac{\sin x}{x}.$$

Methods based on 1) and 3) require a calculator or computer because of the large number of arithmetic operations they involve, and they are called numerical methods. In the next section, we explore a number of numerical methods stemming from the definition of an integral. In Section 10.5, we will examine some methods based on randomizing.

10.2 Standard numerical methods

An approximate value (a *Riemann sum*) for $\int_a^b f(x)\,dx$, is determined by dividing the interval $[a, b]$ into n subintervals (call them $[x_{i-1}, x_i]$, $a = x_0 < x_1 < ... < x_n = b$) and choosing a sampling point in each subinterval (call the sampling point in the ith subinterval c_i). To compute the Riemann sum, you add together the n products

$$f(c_i) \times \text{length of } i^{th} \text{ subinterval}$$

If we let $h_i = x_i - x_{i-1}$, then the Riemann sum is

$$\sum_{i=1}^{n} f(c_i)h_i.$$

We can compute the value of the integral to as many decimal places as needed by computing approximating sums in which the lengths of the subintervals get smaller and smaller (and n gets larger and larger).

The standard numerical methods all depend on subdividing the interval of integration $[a, b]$ into n *equal* pieces, so $h_i = h = (b - a)/n$ for each i (see Figure 10.1).

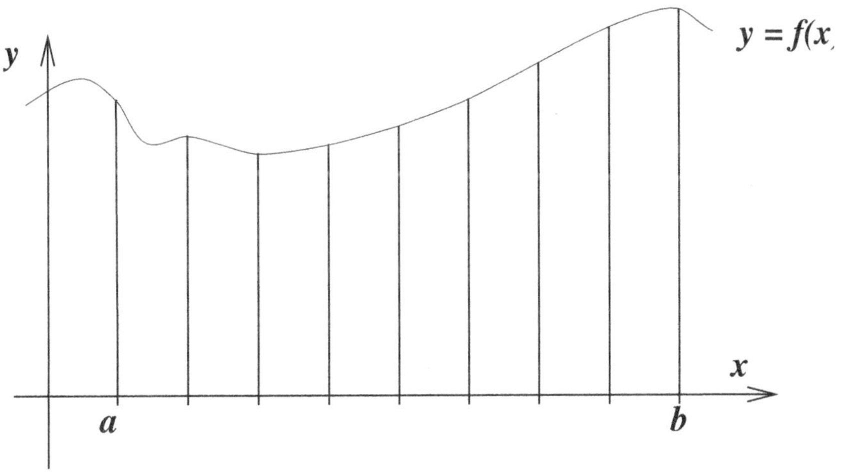

Fig. 10.1: $[a, b]$ divided into equal subintervals

Exercise 1: Explain why, in terms of the notation above, we have $x_i = a + ih$ where $0 \leq i \leq n$. Using this notation, what is x_0? What is x_n?

We will consider five numerical methods. Each corresponds to different choices to approximate the curve $y = f(x)$ up at the top of each "strip" (see Figure 10.2).

The first three methods involve Riemann sums. They approximate the curve by a horizontal line of height $f(c_i)$, producing a *rectangular* strip. In the first case, $c_i = x_{i-1}$, the left endpoint; in the second, $c_i = x_i$, the right endpoint; and in the third, $c_i = (x_{i-1} + x_i)/2$, the midpoint of the subinterval. The fourth method approximates the graph by the line segment from $(x_{i-1}, f(x_{i-1}))$ to $x_i, f(x_i))$, forming a *trapezoidal* strip. The last method, called *Simpson's rule*, requires an even number of subintervals, and then each pair of adjacent subintervals, $[x_{i-1}, x_{i+1}]$ is topped by the parabola passing through the three points $(x_{i-1}, f(x_{i-1}))$, $(x_i, f(x_i))$ and $(x_{i+1}, f(x_{i+1}))$. (To avoid overlaps, we take $i = 1, 3, \dots, n - 1$.)

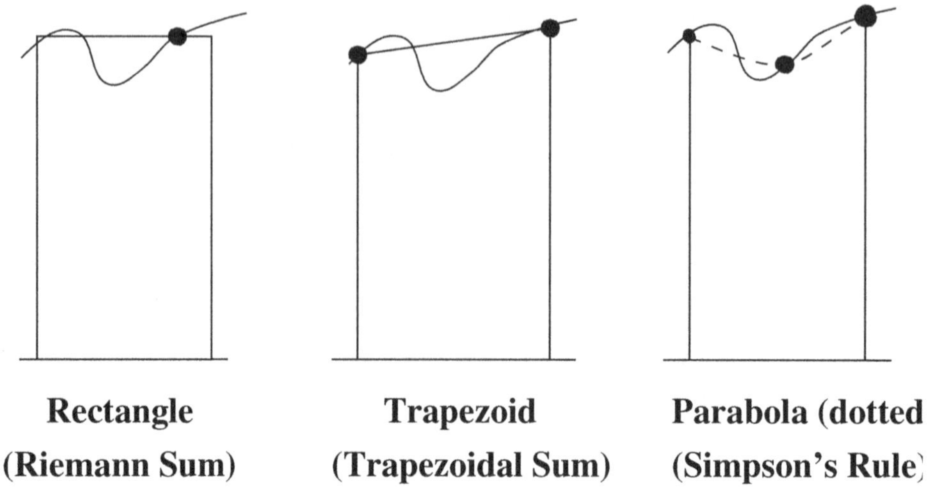

Fig. 10.2: Three approximations

If we use the left endpoints, the approximation we obtain is

$$L_n = \sum_{i=0}^{n-1} f(x_i)h.$$

Exercise 2: Explain why the corresponding sum for the right endpoint approximation is

$$R_n = \sum_{i=1}^{n} f(x_i)h.$$

If $S_n = \sum_{i=1}^{n-1} f(x_i)h$, explain why $L_n = S_n + f(a)h$ and $R_n = S_n + f(b)h$. What is the sum M_n for the midpoint approximation? Recall that the formula for the area of a trapezoid of height h with bases b_1 and b_2 is $[b_1 + b_2]h/2$.

Exercise 3: Explain why the formula for the trapezoidal approximation is

$$T_n = \sum_{i=0}^{n-1} [f(x_i) + f(x_{i+1})]h/2,$$

and explain why T_n is the *average* of L_n and R_n.

Exercise 4: For distinct numbers a, b, and c, a unique parabola passes through the three points (a, A), (b, B) and (c, C). Verify that the parabola whose equation is $y = q(x)$ for

$$q(x) = A\frac{(x-b)(x-c)}{(a-b)(a-c)} + B\frac{(x-a)(x-c)}{(b-a)(b-c)} + C\frac{(x-a)(x-b)}{(c-a)(c-b)}$$

passes through these points. (How do you know that the graph of $y = q(x)$ actually *is* a parabola?)

Exercise 5: Consider the special case where $a = -h$, $A = f(a)$; $b = 0$, $B = f(b)$, and $c = h$, $C = f(c)$. Draw a picture to illustrate this case. Use calculus to find a formula (in terms of h, $f(-h)$, $f(0)$ and $f(h)$) for the area under the graph of $y = q(x)$ on the interval $[-h, h]$.

Exercise 6: For Simpson's rule we approximate the curve $y = f(x)$ on an adjacent pair of subintervals by a parabola passing through the three consecutive points

$$(x_{i-1}, f(x_{i-1})),\ (x_i, f(x_i)),\ (x_{i+1}, f(x_{i+1}))$$

What is the formula for $q(x)$ in this case? Use your answers to Exercises 5 and 6 to explain why the area under the graph of $q(x)$ between x_{i-1} and x_{i+1} is

$$\frac{1}{3}\left\{f(x_{i-1}) + 4f(x_i) + f(x_{i+1})\right\} h.$$

Exercise 7: Assume n is even. Explain why the approximating sum for Simpson's rule is

$$\begin{aligned} S_n =& \frac{1}{3}\{f(x_0) + 4f(x_1) + 2f(x_2) + 4f(x_3) + 2f(x_4) + ... \\ &+ 2f(x_{n-2}) + 4f(x_{n-1}) + f(x_n)\} h. \end{aligned}$$

Note: In the next section you will see that there is another way to think of Simpson's rule by relating it to the midpoint and trapezoidal approximations.

10.3 Automating the standard methods

It is not hard to automate the calculation of the five approximating sums in the previous section. Here is pseudocode for a program INTEGRAL that calculates all five. In order to make the programs simpler, we will choose $n = 2m$ to be an *even* integer. For the first four approximations, we will think of our subintervals as $a = x_0 < x_2 < x_4 < ... < x_{2m} = b$. Then the midpoints of the subintervals are $x_1, x_3, ..., x_{2m-1}$. Thus, the program will calculate the approximating sums L_m, R_m, M_m, T_m and S_n. The pseudocode assumes that the programming language used starts each array index at 1. Because we begin our sequence at x_0, our array indices will be one greater than their corresponding x_i index.

Program outline: Itegral

```
Input: a function f, an interval [a,b], and an integer m > 0
Output: the value of the left endpoint, right endpoint, mid-
        point, trapezoidal and Simpson's approximating sums:
        L(m), R(m), M(m), T(m), S(2*m)

%Calculate an array, Fval(i), of function values, i = 0 to n
h := (b-a)/n    %Length of (half) a subinterval
INITIALIZE fval as a 2m dimenstional vector
x := a          %Initialize x
FOR i = 0 TO n
        fval(i) = f(x)
        x := x + h
END FOR

%Calculate  "master sum" for left, right, and trap sums
fsum := 0
FOR i = 1 TO (m-1)
        fsum := fsum + fval(2 * i + 1)
END FOR

Left := [fsum + fval(1)] * (2 * h)
```

```
Right := [sum + fval(2 * m + 1)] * (2 * h)
Trap := [Left + Right]/2

%Calculate midpoint sum
fsum2 := 0
FOR i = 1 TO m
        fsum2 := fsum2 + fval(2 * i)
END FOR
Mid :=  sum * (2 * h)

%Calculate Simpson's rule sum
Simp :=  [Fval(1) + 4 * fsum2 + 2 * fsum + fval(2n+1)] * h/3

OUTPUT Left, Right, Mid, Trap, Simp
```

For some of the questions in the next section, you will use the program INTEGRAL to compare the accuracy of the different methods in the case when the actual value of the integral is known by the fundamental theorem. In these cases, what is of interest is not the value of the approximating sums, but the *difference* between each approximation and the actual value—the error in the approximation. For this task, a slight variant, INTEGRAL-2, is useful; it merely adds the actual value of the integral as an input at the start and prints out the differences between the actual and estimated values.

Program outline: Itegral-2

```
Input: a function f, interval [a,b], integer m, and the
        actual value A of the integral
Output: the differences between the leftpoint, rightpoint,
        midpoint, trapezoidal and Simpson's approximating
         sums and the actual value:
        L(m) - A, R(m) - A, M(m) - A, T(m) - A, S(2*m) - A

RUN Integral code

OUTPUT Left - A, Right - A, Mid - A, Trap - A, Simp - A
```

Exercise 8: Explain why the program INTEGRAL does what it is supposed to do.

10.4 Questions to explore

Question 1: Compare the accuracy of the five methods just given by applying them to some examples where you know the answer. In each case, calculate the approximation and the error = approximation − actual value.

(a) $f(x) = \cos x, \quad [a,b] = [0, \frac{\pi}{2}]$.

(b) $f(x) = 2x + 1, \quad [a,b] = [0,1]$.

(c) $f(x) = 4 - x^2, \quad [a,b] = [0,2]$.

(d) $f(x) = 5x^3 - 6x^2 + .3x, \quad [a,b] = [-1,3]$.

Compare the size of the error for each of the methods for the same choice of $n = 2m$, and also observe the effect of doubling n on the size of the error for each method. You might start with $n = 2, 4, 8, 16, \ldots$. It is sometimes helpful to keep track of the improvement in your approximations by computing the ratio $\text{Error}(n)/\text{Error}(2n)$ for the various methods, where $\text{Error}(n)$ means the error in the approximating sum when you use n subintervals.

Question 2: If you choose $f(x) = \sqrt{|1-x^2|}$, $[a,b] = [-1,1]$, you will obtain an approximation to $\frac{\pi}{2}$ (Why?). If you were to double m each time, how large a value of m, for each method, do you need to obtain π accurate to 2,3,4 etc. decimal places? Also try $m = 10, 100, 1000$. [Note: the absolute value bars are needed because for large m, round-off error can produce negative values of $1 - x^2$ for x near ± 1.]

Question 3: Choose other examples of functions and intervals on which to apply the five methods. (The integral $\int_0^1 dx/(1+x^2)$ is quite interesting!)

Question 4: How do the midpoint and trapezoidal approximations compare? (Are you surprised?) To get some insight into this comparison, sketch a typical "strip" with its midpoint rectangle, and create a trapezoid we'll call the *midpoint trapezoid.* Do this by rotating the horizontal line through (midpoint, f(midpoint)) about this point until it is tangent to the graph of $y = f(x)$. How do the areas of the midpoint rectangle and the midpoint trapezoid compare?

Now, draw some pictures comparing the midpoint trapezoid with the trapezoid used in the trapezoidal approximation. Try comparing the standard and midpoint trapezoids for several, variously shaped, graphs. Which trapezoid looks like its area is closer to the area under the curve? What do your pictures suggest about the relationship between the *signs* of the errors for the midpoint and trapezoidal approximations?

Question 5: Your results in Questions 1–4 should suggest that the errors in the midpoint and trapezoidal approximations have opposite signs, and that the error in the midpoint sum is about half that in the trapezoidal sum. Based on this observation, explain why forming the *weighted* sum

$$W_m = \frac{1}{3}\left[2 \times M_m + 1 \times T_m\right]$$

should give a very good approximation to the integral. Use your results from Questions 1–3 to calculate some values of W_m for $m = 1, 2, 4, \ldots$. What do you get? How do your results compare to the results S_{2m} using Simpson's rule?

Question 6: Go back to the special case in Exercise 5, where you looked at the graph of $y = f(x)$ on the interval $[-h, h]$. Calculate M_1, T_1, and W_1 on this interval, and calculate S_2 on the pair of adjacent subintervals $[-h, 0]$ and $[0, h]$. Your answers will all be in terms of h and $f(-h)$, $f(0)$ and $f(h)$. What do you find?

Question 7: Use your answers to Questions 5 and 6 to guess a formula for S_{2m} in terms of M_m and T_m. Can you prove your formula is correct?

Question 8: If $f(x)$ is linear, what do you notice about the midpoint, trapezoidal and Simpson's rule approximations to the integral of f? If $f(x)$ is quadratic ($f(x) = ax^2 + bx + c$, some a, b, c) what do you notice about the Simpson's rule approximation? Can you explain these observations?

Question 9: Now try some cubic polynomials $f(x) = ax^3 + bx^2 + cx + d$ using Simpson's rule. What do you notice? To see what's going on, compute *algebraically* both the Simpson's rule approximation and the actual integral in the special case of Exercise 5 ($n = 2$ and working on the interval $[-h, h]$) for $f(x) = ax^3 + bx^2 + cx + d$.

Question 10: Now choose a function we *don't* know how to integrate, say $f(x) = e^{-x^2}$, on $[a, b] = [-1, 1]$. Compare the approximations to this integral produced by the various methods.

Question 11: What about $f(x) = \sin x/x$ on $[a, b] = [1, 2]$? or on $[a, b] = [0, 1]$? (There are problems in the latter case near $x = 0$. What can you do to avoid them?).

Integrals describing *arc length* occur naturally, but we can seldom evaluate them analytically, so they are natural candidates for our numerical methods. The "circumference" of an ellipse is such an integral.

Consider the ellipse $x = a\cos t$, $y = b\sin t$, for $0 \le t \le 2\pi$. (Notice that $x^2/a^2 + y^2/b^2 = 1$). Assume that the *major axis* of the ellipse is horizontal, so that $b \le a$. The *speed* of the parametrization is

$$\sqrt{(x'(t))^2 + (y'(t))^2)} = \sqrt{a^2 \sin^2 t + b^2 \cos^2 t},$$

and the arc length $L =$ distance traveled $= \int_0^{2\pi}$ speed dt, so

$$\begin{aligned} L &= \int_0^{2\pi} \left(a^2 \sin^2 t + b^2 \cos^2 t\right)^{\frac{1}{2}} \, dt \\ &= 4 \int_0^{\frac{\pi}{2}} \left(a^2 \sin^2 t + b^2 \cos^2 t\right)^{\frac{1}{2}} \, dt \\ &= 4a \int_0^{\frac{\pi}{2}} \left(1 - \left(1 - \frac{b^2}{a^2}\right) \cos^2 t\right)^{\frac{1}{2}} \, dt. \end{aligned}$$

(Be sure you can account for each step in the calculation above.)

It is conventional to give the name $k^2 = (1 - b^2/a^2)$ to the expression appearing in the final version of the formula for L. (How do we know $(1-b^2/a^2)$ is non-negative?) Then the non-negative root k is called the *eccentricity* of the ellipse.

Question 12:

(a) How large can k be? How small? What is the eccentricity of a circle?

(b) The integral $\int_0^{\pi/2}(1 - k^2\cos^2 t)^{\frac{1}{2}}\,dt$ is called a *complete elliptic integral.* Choose $a = 1$ so $k^2 = 1 - b^2$. Use Simpson's rule to find the values of this integral for $k = 0, 0.1, 0.2, \ldots, 0.9, 1.0$. How do these results correspond to your intuition about lengths of the various ellipses?

10.5 Monte Carlo methods

We can use a pile of stones to measure the area of a pond as follows. Suppose the pond lies in a rectangular field of known area (see Figure 10.3). Throw stones at random so that they all land within the field and are equally likely to land in any given location in the rectangle. A reasonable estimate of the pond's area could be the area of the field times the fraction of stones which make a splash. (Why is this so?) This suggests how we can determine integrals approximately using chance.

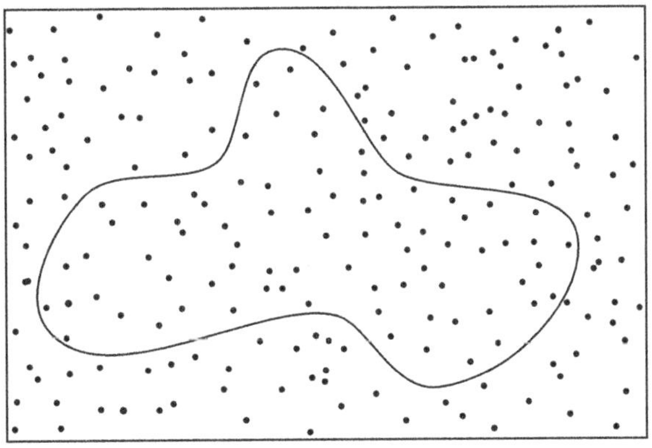

Fig. 10.3: A pond

To make life easy, suppose that we have a function $f(x)$ defined on an interval $[a,b]$ with $0 \leq f(x) \leq H$ for $a \leq x \leq b$, where H is some fixed nonnegative number (see Figure 10.4). Suppose also that we randomly select m pairs of numbers x_i, y_i, $1 \leq i \leq m$, with $a \leq x_i \leq b$, and $0 \leq y_i \leq H$. For each i, we record a "splash" if $f(x_i) - y_i \geq 0$, and let s be the number of splashes. An approximation to our integral is then

$$H \times (b-a) \times \frac{s}{m}.$$

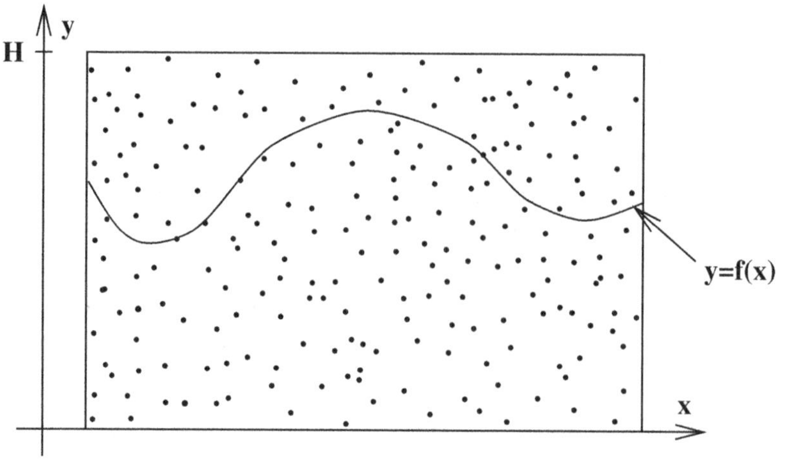

Fig. 10.4: A "pond" beneath a graph

Exercise 9: Explain the analogy between this procedure and measuring the area of the lake by tossing stones.

An alternative method is based on the fact that the *average value* of $f(x)$ on the interval [a,b] is

$$\frac{1}{b-a}\int_a^b f(x)dx.$$

If we now randomly select m points x_i with $1 \leq i \leq m$ in the interval $[a,b]$, we can approximate the average value of f by

$$\frac{1}{m}\sum_{i=1}^{m} f(x_i).$$

Hence an approximation to the integral is

$$\frac{(b-a)}{m}\sum_{i=1}^{m} f(x_i).$$

Both methods are referred to as *Monte Carlo* techniques (although using chance mechanisms to solve mathematical problems isn't as expensive as visiting the gaming tables in Monte Carlo).

Exercise 10: Explain the logic of the second method.

We can use a computer program to simulate our two Monte Carlo methods. Most computer languages have a function *rand* that selects a random number between 0 and 1. The two programs MONTE-1 and MONTE-2 below exploit this function.

Program outline: Monte-1

```
Input: a positive integer m, a function f(x), real numbers
a, b, H, where  0 <= f(x) <= H for a <= x <= b.
Output: an estimate for the integral of  f  on  [a,b]
using method 1 ("throwing stones")

Reset seed for random number generator
s = 0    ! Set counter for s = number of "splashes"
FOR i = 1 TO m
x := a + (b-a) * rand  %Select x "randomly" in [a,b]
y := H * rand          %Select y "randomly" in [0,H]
IF (f(x) - y) >= 0
s:= s + 1
END IF
END FOR
est := H * (b-a) * s/m    %estimate of integral
OUTPUT est
```

Exercise 11: Explain why both programs work.

Question 13: Try both methods out on some of our previous integrals. (E.g., $f(x) = \cos x$, for $0 \leq x \leq \pi/2$.) How do the Monte Carlo methods compare in accuracy to our earlier methods?

Program outline: Monte-2

```
Input: a positive integer m, a function f(x), real numbers
    a, b
Output: an estimate for the integral of  f  on  [a,b]
       using method 2 (based on the average value of f)
sum := 0
FOR i = 1 TO m
   x := a + (b-a) * rand  %Select x "randomly" in [a,b]
   sum := sum + f(x)
END FOR
est :=   (b-a) * sum/m     %estimate of integral
OUPUT est
```

Question 14: How do the Monte Carlo methods depend on the number m of points randomly chosen? (This is a very open ended problem! In order to make any headway, you will have to try some *large* values of m; try $m = 5000, 10000, \ldots$.)

10.6 Higher dimensions

Simpson's rule is, generally, a better method of approximating integrals than either of the Monte Carlo methods. But this is only true for functions of a single variable. When we integrate functions of $2, 3, 4, \ldots$ variables, the Monte Carlo methods become much more attractive (and for $n \geq 3$ better) than the analogue of Simpson's rule.

Even if you have not studied multivariable calculus, you can investigate higher dimensional integrals using Monte Carlo methods. For example, using the second method, you can estimate the integral of

$$f(x, y) = x^2 + 6xy + y^2$$

over the unit square in the x-y plane $0 \leq x,\ y \leq 1$, which we will call A. Since the area of A is 1, the average value of f on A is

$$\int\int_A f(x,y)\,dxdy,$$

and this integral can be approximated by

$$\frac{1}{m}\sum_{i=1}^{m} f(x_i, y_i)$$

where the $0 \leq x_i,\ y_i \leq 1$ are $2m$ randomly selected numbers.

We can tackle an interesting geometric problem using a variant of our first Monte Carlo method, counting "splashes." Imagine a ball of radius 1/2 sitting inside the unit cube centered at the origin. Figure 10.5 shows the ball for dimension 3, but the mathematics makes sense in dimension n for *any* positive integer n. In the case where $n = 1$, the "volume" of the "ball" is simply the length of the line segment $[-1/2, 1/2]$. Similarly, "volume" in dimension 2 means the area of the circle of radius 1/2.

In dimension n, the *unit cube* centered at the origin consists of all n-tuples $(x_1, x_2, \dots, x_n)$ with $-1/2 \leq x_i \leq 1/2$ for each i. The *sphere* of radius 1/2 consists of all n-tuples with

$$\sqrt{x_1^2 + x_2^2 + \cdots + x_n^2} \;\leq\; \frac{1}{2}.$$

Question 15: What is the value of the quotient

$$\frac{\text{volume ball}}{\text{volume cube}} = \frac{\text{volume ball}}{1},$$

and how does this value behave as the dimension n increases?

You can attack this problem using the program BALL. Pseudocode for the program appears below, and working code in *MATLAB/Octave* is at the end of this chapter. In the program BALL, n is the dimension and m is number of points chosen randomly inside the unit cube. Picking a point

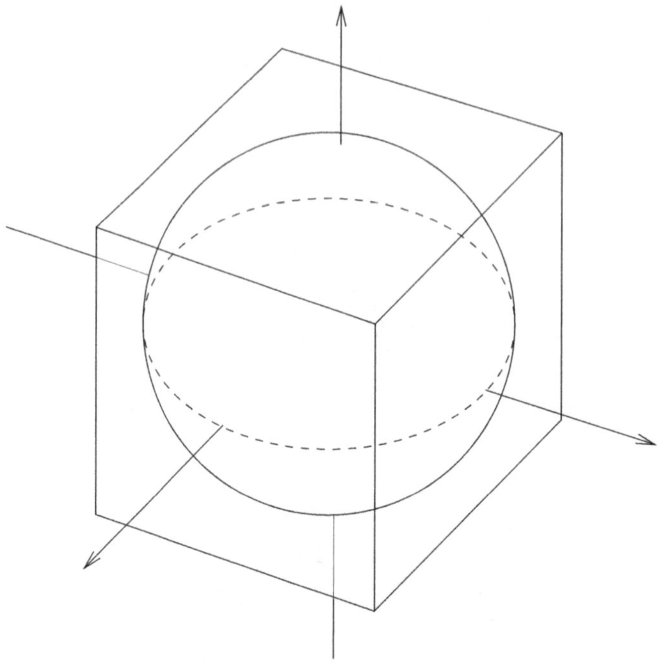

Fig. 10.5: Unit ball

inside the cube means picking n coordinates between $-1/2$ and $1/2$. Since the function *rnd* returns a random number between 0 and 1, you have to subtract off $1/2$ to get a number between $-1/2$ and $1/2$. The chosen point will be inside the ball of radius $1/2$ if and only if the sum of the squares of the coordinates is less than or equal to $1/4$. (Do you see why?) As in our program MONTE-1, s is the number of randomly chosen points that land *inside* the ball. Since the volume of the unit cube is 1, the fraction s/m is an estimate of the ratio volume ball/volume cube and thus an estimate of the volume of the ball of radius $1/2$ in dimension n.

Program outline: Ball

```
Input: positive integers n and m ! n = dimension, m = #
       points
Output: estimate of the volume of the ball of radius 1/2
s := 0  %Counts "splashes"
FOR i = 1 TO m
   sum := 0
   FOR j = 1 TO n
       x = rnd
       sum = sum + (x - 1/2)^2
   END FOR
   IF sum <= 1/4
       s = s + 1
   END IF
END FOR
OUTPUT s/m
```

10.7 Computer programs

10.7.1 *MATLAB/Octave* programs

Program: Integral

```
function [left,right,mid,trap,simp]=integral(f,a,b,m)
%function [left,right,mid,trap,simp]=integral(f,a,b,m)
%
%Gives numerical estimates of the integral of a function on the interval [a,b]
%
%Inputs
%  f - function
%  a - left endpoint of integral
%  b - right point of integral
%  m - number of subintervals
```

```
%Output
%  left - left point method
%  right - right point method
%  mid - mid point method
%  trap - trapezoid method
%  simp - Simpson's method


fval=zeros(2*m+1,1); %vector to hold set of function values
h=(b-a)/(2*m); %subinterval width
x=a; %first x value
for i=1:(2*m+1)
fval(i)=f(x);  %evaluate function
x=x+h; %next x value
end; %for

fsum=0; %variable for sum of function values

%sum function values for left right and trapezoid
for i = 1:m-1
fsum=fsum+fval(2*i+1);
end; %for
left=(fsum+fval(1))*2*h;
right=(fsum+fval(2*m+1))*2*h;
trap=(left+right)/2;


%sum function values for midpoint method
fsum2=0;
for i = 1:m
fsum2=fsum2+fval(2*i);
end; %for
mid=fsum2*2*h;


%Simpson's method
simp=(fval(1)+4*fsum2+2*fsum+fval(2*m+1))*h/3;
```

This program requires the input of the function to be integrated. The easiest way to do this is to define an anonymous function. For example, if we want to estimate the integral of $\sqrt{x^2+1}$ with $[a,b]=[0,2]$ and $m=4$, enter:

```
>> f = @(x) sqrt(x^2 +1)
>> [left,right,mid,trap,simp]=integral(f,0,2,4)
```

The function needs to be defined in the same way for the next program.

Program: Integral-2

```
function [lefterr,righterr,miderr,traperr,simperr]=integral2(f,a,b,m,A)
%function [lefterr,righterr,miderr,traperr,simperr]=integral2(f,a,b,m,A)
%
%Gives numerical estimates of the integral of a function on the
%interval [a,b]
%
%Inputs
%  f - function
%  a - left endpoint of integral
%  b - right point of integral
%  m - number of subintervals
%  A - actual exact value of integral
%Output
%  lefterr - error left point method
%  righterr - error right point method
%  miderr - error mid point method
%  traperr - error trapezoid method
%  simperr - error Simpson's method

[left,right,mid,trap,simp]=integral(f,a,b,m);

lefterr = left-A;
righterr = right-A;
miderr = mid-A;
traperr = trap-A;
simperr = simp-A;
```

Program: Monte-1

```
function est=monte1(f,a,b,m,H)
%function est=monte1(f,a,b,m,H)
%
%Gives monte carlo estimate of the integral of a function on the interval
%[a,b] using "stone throwing" methodbased on areas
%
%Inputs
%  f - a positive function`
%  a - left endpoint of integral
%  b - right point of integral
%  m - function evaluations
%  H - least upper bound of f on interval [a,b]
%Output
%  est - integral estimate

s = 0; %set counter for number of splashes to 0
for i=1:m
x=a+(b-a)*rand;
y=H*rand;
if y<=f(x)
s=s+1;
end; %if
end; %for
est=H*(b-a)*s/m;
```

Program: Monte-2

```
function est=monte2(f,a,b,m,H)
%function est=monte2(f,a,b,m,H)
%
%Gives monte carlo estimate of the integral of a function on the
%interval [a,b] based on average value
```

```
%
%Inputs
%  f - function
%  a - left endpoint of integral
%  b - right point of integral
%  m - number of function evaluations
%Output
%  est - integral estimate

fsum = 0; %variable for sum of function values
for i=1:m
x=a+(b-a)*rand;
fsum=fsum+f(x);
end; %for
est=(b-a)*fsum/m;
```

Program: Ball

```
function est=ball(n,m)
%function est=ball(n,m)
%
%Gives monte carlo estimate of the volume of a n-dimensional ball
%of radius 1/2
%
%Inputs
%  n - dimemsion of the ball
%  m - number of points
%Output
%  est - integral estimate

s=0; %count of number of splashes
for i=1:m
x=rand(1,n); %a 1xn vector of random numbers
rsq=sum((x-0.5).^2); %the radius squared
if rsq<1/4
s=s+1; %increment splash counter
```

```
end; %if
end; %for
est=s/m;
```

Chapter 11

Sequences and Series

11.1 Introduction

You have certainly seen examples such as

$$0.3, 0.33, 0.333, 0.3333 \ldots \to \frac{1}{3}$$

where we speak of an infinite sequence of numbers *converging* to a number. Similarly, you have probably seen examples where we add the members of an infinite sequence of numbers to form an infinite series, and speak of such a series *converging* to a sum. In fact, the sequence of numbers above can be rewritten as a series

$$\frac{3}{10} + \frac{3}{10^2} + \frac{3}{10^3} + \frac{3}{10^4} + \ldots = \frac{1}{3}.$$

The notion of convergence of a sequnce of numbers, and the related notion of convergence of a series, is central to mathematical analysis. In this chapter, you will

- Explore, both informally and formally, the convergence of certain sequences and series, and

- Examine in some detail the harmonic series and its divergence.

You will also use the computer program SEQUENCE AND SERIES. This program allows you to view, both graphically and numerically, many values of a sequence whose terms you can enter in closed form. It also allows you see, again both graphically and numerically, the corresponding series formed by adding members of the sequence. The graphical output allows you to "see" the convergence (or lack thereof).

11.2 The mathematical ideas

In this lab, we will be interested in ways that a **sequence** $\{a_k\}$ of numbers,

$$a_1, a_2, \dots, a_k, \dots$$

may *converge* to a limit L, and ways in which a sequence may be used to build a **series** of numbers

$$\sum_{k=1}^{\infty} a_k = a_1 + a_2 + \dots$$

that may itself *converge* to a sum S. The symbol k is often called the *index* for the series, and it may have its first value at 1 or 0 or any other finite integer. Infinite series are frequently found in applications in mathematics, physics and the other sciences.

Informally, we say that a sequence a_k **converges** to a limit L if the terms of the sequence get arbitrarily close to L as k gets arbitrarily large (approaches positive infinity).

Exercise 1: This is a somewhat philosophical exercise. Write a paragraph describing what this informal definition of convergence of a sequence $\{a_k\}$ to a limit L means to you. Be as precise as you can. Articulate what it means for the numbers $\{a_k\}$ to be "arbitrarily close" to L and for k to get "arbitrarily large". Use common sense, and be as mathematical as you can.

A convergent infinite series may be thought of as the limit (in the sense that you just characterized) of the sequence $\{s_n\}$ of **partial sums**

$$s_n = \sum_{k=1}^{n} a_k = a_1 + a_2 + ... + a_n,$$

and we will say that the **series converges** to a real number S provided the sequence $\{s_n\}$ converges to S as n approaches infinity. If the series converges to S, then we say that S is its **sum**. We write

$$S = \sum_{k=1}^{\infty} a_k = a_1 + a_2 + \ldots.$$

If a series does not converge, it will be said to **diverge**.

A pictorial example

There are many ways to examine a series, and these often allow us to say whether the series converges or it diverges. Sometimes, in the case of convergence, we can also identify the numerical sum of the series. Frequently there is a nice interplay between the numerical series and a geometric representation.

For instance, suppose you have the series:

$$\sum_{k=1}^{\infty} \frac{1}{2^k} = \frac{1}{2} + \frac{1}{4} + \frac{1}{8} + ...$$

If you have seen the theory of *geometric series*, you will know that this series converges, and you will know what its sum is. On the other hand, you can arrive at the same conclusion by thinking of rectangles of area $1/2, 1/4, 1/8, ...$ filling a 1-unit by 1-unit square when they are arranged as in Figure 11.1

Clearly, the sum of these areas is 1 square unit. On the other hand, if we *combine* these same rectangles pairwise into shapes such as in Figure 11.2 we also get a total area of one square unit.

Exercise 2: What numerical series, summing to 1, does the diagram in Figure 11.2 suggest to you?

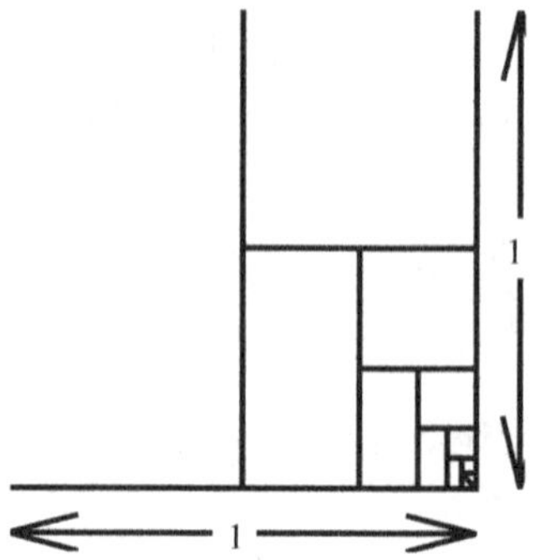

Fig. 11.1: Rectangles filling a unit square

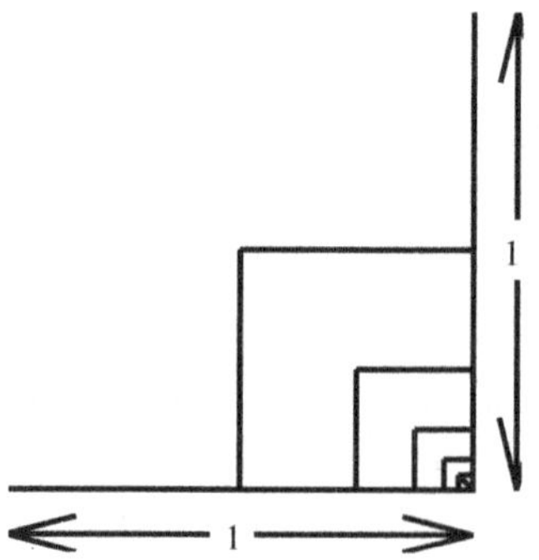

Fig. 11.2: Another way to fill the unit square

An algebraic example

So far we've looked geometrically at the convergence of a series. Suppose, now, we take a quick look at the series

$$\sum_{k=1}^{\infty} \frac{1}{k(k+1)} = \frac{1}{1 \cdot 2} + \frac{1}{2 \cdot 3} + \ldots$$

Exercise 3: Find an expression for the partial sum

$$\sum_{k=1}^{n} \frac{1}{k(k+1)}$$

and find the sum of the series.

[Hint: Note that the expression $\frac{1}{k(k+1)}$ can be otherwise written $\frac{1}{k} - \frac{1}{k+1}$. You can prove this algebraically by combining these two fractions over a common denominator and simplifying, or, if you have seen it in calculus, you can use the method of "partial fractions" on the initial expression.]

11.3 The harmonic series

We now begin an examination of the series formed from the sequence whose terms are $a_k = 1/k$. Our series, then, is

$$\sum_{k=1}^{\infty} \frac{1}{k} = 1 + \frac{1}{2} + \frac{1}{3} + \dots$$

If you have studied infinite series in calculus, you probably know that this series, called the *harmonic series*, diverges. That is, its sequence of partial sums does not converge to a real number and, in fact, it increases monotonically without bound. A question of interest is: *how quickly, or slowly, does it diverge (in this case, get large)*?

Exercise 4: By hand or with a calculator, find s_2, s_3, s_4, s_5, and s_6 for the harmonic series.

As a way to get a feeling for how fast the series grows, define the function $J(n)$ to be the least integer greater than or equal the partial sum

$$s_n = \frac{1}{1} + \frac{1}{2} + \frac{1}{3} + \dots + \frac{1}{n}.$$

So, for example, $J(1) = 1$, $J(3) = 2$, $J(4) = 3$.

Exercise 5: By hand or with a calculator, find $J(5)$, $J(6)$, and $J(11)$.

Figuring out $J(n)$ by hand quickly becomes tedious. You can use the program SEQUENCE AND SERIES below to explore the behavior for large n of s_n and $J(n)$ for the harmonic series. Before turning to that section, try to answer the following questions. If you cannot, keep them in mind as you explore further.

Exercise 6: Suppose you know $J(n)$. Can you estimate $J(2n)$? In particular, can you say that $J(2n) > J(n)$. Why or why not? Given n, can you find N as a function of n which guarantees that $J(N) = J(n) + 1$? (Work a few examples for small values n. Can you make any conjectures?)

In this lab, you will shortly plot the first 10, 100, 500, and 1000 terms of the harmonic sequence along with the corresponding partial sums. The program SEQUENCE AND SERIES can be used to accomplish this task.

Program outline: Sequence and Series

```
Input: a function f that gives the terms (f(k)=a_k),
        a positive integer n
Output: the sequence of terms a_k, the sequence of partial
        sums s_k, and their plots for (1<=k<=n)

INITIALIZE ak as an n-dimensional vector
INITIALIZE sk as an n-dimensional vector

ak(1) := f(1)
sk(1) := ak(1)

FOR k = 2 TO n
  ak(k) := f(k)
  sk(k) := sk(k-1) + ak(k)
END FOR

PLOT sk and pk vs k
OUTPUT sk pk
```

Figure 11.3 shows the output of SEQUENCE AND SERIES for the harmonic series with $n = 500$. Notice that the plot shows individual points for the first few k values, but then the plot appears to be a continuous line. In reality the terms of a sequence and the partial sums will always be discrete points, it is just that they are so close together they appear as a continuous line.

The partial sums are always increasing. Why is this? Will the partial sums increase for every series? The function $J(n)$ can be found by simply rounding up the last entry in the sk vector output by the program.

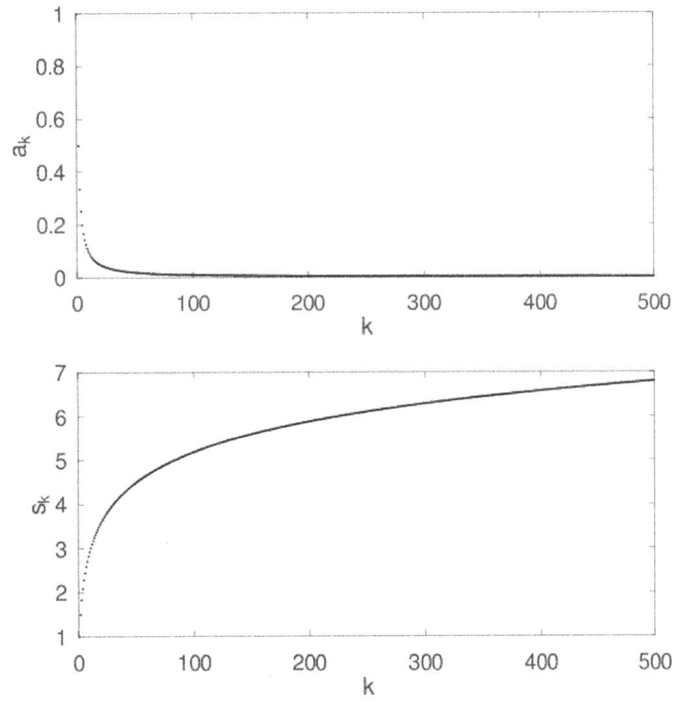

Fig. 11.3: The first 500 terms (above) and partial sums (below) of the harmonic series.

Try the following exercises regarding $J(k)$.

Exercise 7: Summarize in a chart like Table 1 the results that you got in the previous section (showing values of k, of a_k, the sum for k terms, and $J(k)$).

Exercise 8: Find $J(500)$ and $J(1000)$. Fill in this information on your chart from Exercise 7.

Exercise 9: Try the same technique for five or six other values of your choice for k between 100 and 1000, extending the chart to record your values at each stage.

k	a_k	Sum	$J(k)$
3	0.3333...	1.83333...	2
4	0.2500...		
10	0.1000...		
100			

Table 1

Exercise 10: First, we turn the question from the previous exercises around. This time, for each of the integers $1, 2, 3, 4, 5, 6, 7, 8, 9$, find the *largest* values of n that give these integers as $J(n)$. For example, for the integer 4, the value you are looking for is 30. This is the case because the SUM for $n = 30$ is 3.99498 ... (so $J(30) = 4$), whereas the SUM for $n = 31$ is 4.02724... ($J(31) = 5$).

Record these in a chart like Table 2 (ignoring for now the column that says "Ratios"):

n	Sum(n)	Sum(n+1)	$J(n)$	Ratios
1	1.. ...	1.50000..	1	xxxxxxxx
3	1.83333..	2.08333333	2	3.00000
			3	
30	3.99498...	4.02724520	4	
			5	
			6	
			7	
			8	
			9	

Table 2

Exercise 11: In the Ratios column of Table 2, fill in numerical ratios of the *current row's* n to the *previous row's* n. So, for instance, the 3.00000 that is already filled in is the ratio of the values 3 to 1 in the n's column. What do you notice about the ratios? Would you care to make any speculations here?

Exercise 12: For this question you will need determine how many terms of the harmonic series can compute in a certain amount of time. Luckily we can use the computer's clock, so we don't need a stopwatch. Instead we can use the program STOPWATCH, which reports how many of terms of the harmonic series the computer has summed after one minute. Assuming that the computer will continue to calculate the same number of terms per minute, estimate how many terms, N, your computer would calculate by this time tomorrow if it were allowed to continue running. [Note: there is no single correct answer to this estimate, since different computers do their work at different speeds, depending on the configuration of the hardware.]

Program outline: Stopwatch

```
Input:  none
Output: the number of terms of the harmonic series summed by
        the computer after each six-second interval up to
        one minute

INITIALIZE Timer
sk := 0 %the partial sum so far
k := 0

FOR i = 1 TO 10
   WHILE (Time Elapsed) < (6 * i)
      k := k+1
      sk := sk + 1/k
   END WHILE
   %Output time & number of terms computed
   PRINT Time Elapsed, k
END FOR
```

11.4 The logarithmic integral

We now attempt an estimate of the sum S, as it will have been calculated by this same time tomorrow—an estimate done without having to leave the computer running all that time! We have just estimated the number N of terms that will be calculated. Next we estimate the sum S for this value of N. Once this is done, $J(N)$ can easily be determined—it's just the first integer greater than or equal to S. We approach this problem with a strategy you may not have predicted.

Look at the function $f(x) = 1/x$ for $1 \leq x \leq 5$. We will partition the interval $[1, 5]$ into 4 equal subintervals (each of length 1 unit). We make the important observation that the 4th partial sum of the harmonic series can be viewed as the sum of the areas of the 4 rectangles pictured in Figure 11.4. Do you see why this is so? Be sure you understand this fully before you go on.

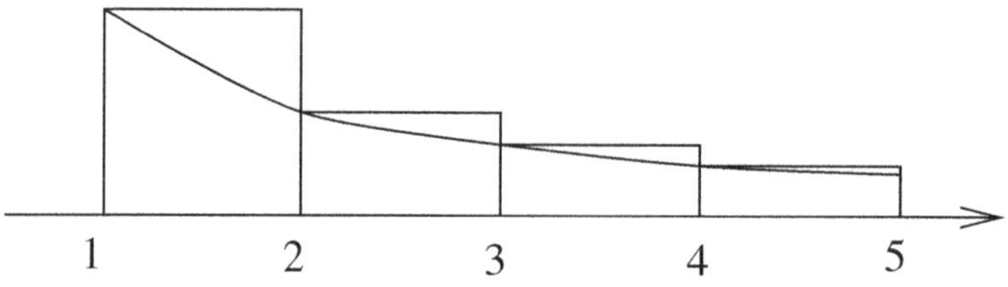

Fig. 11.4: The 4th partial sum of the harmonic series s_4 represented with 4 rectangles

The exact value of the area we have looked at under the curve $y = 1/x$ from $x = 1$ to $x = 5$ is obtainable from the fundamental theorem of calculus. Since $\ln x$ is an antiderivative of $1/x$,

$$\int_1^5 \frac{1}{x}\, dx = \ln 5 - \ln 1 = \ln 5\,.$$

The integral of $1/x$ is called the *logarithmic integral.*

Observe that for any positive integer n

$$\ln(n+1) < 1 + \frac{1}{2} + \frac{1}{3} + ... + \frac{1}{n}$$

We rewrite this:

$$\ln(n+1) < s_n$$

Exercise 13: Now we try some geometry. For any n, the value $s_n - \ln(n+1)$, as represented by the total area of the "triangular" regions above the curve $1/x$ of Figure 11.4, can be made to fit into the unit square labeled U (see Figure 11.5).

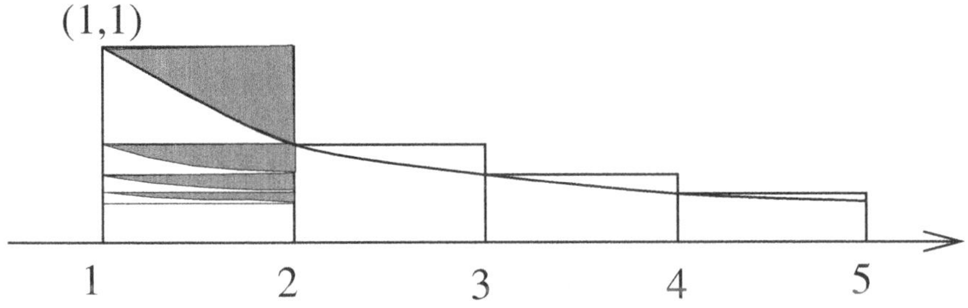

Fig. 11.5: Triangle-like regions moved into the unit square U

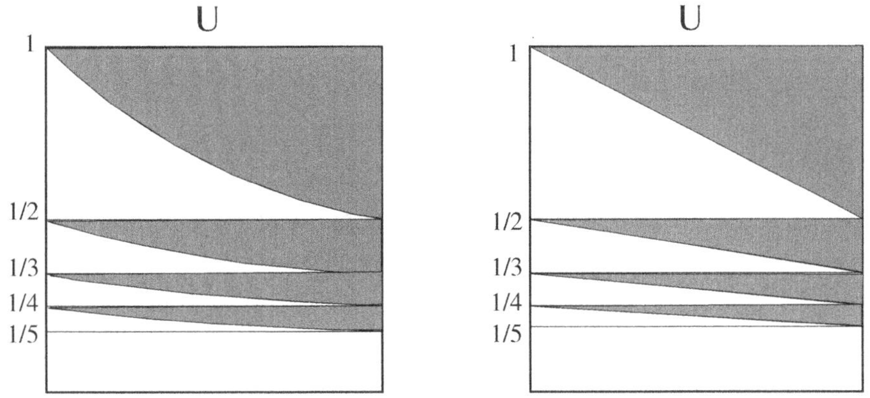

Fig. 11.6: Triangle-like regions in unit square U and actual triangles replacing them

To accomplish this, imagine picking up each of the "triangular" regions within the rectangles lying above $1/x$ between 1 and $n+1$, and moving it horizontally to the left until it just fits inside of U. Then $s_n - \ln(n+1)$, represents the total area that is shaded in U (why is this?). In the left-hand diagram in Figure 11.6, we isolate the square U after the "triangular" regions have been moved.

An n increases, we see that the shaded area increases (since we are adding additional regions to it) so, in particular, $s_n - \ln(n+1)$ is an increasing sequence. Since the shaded region fits into U, we clearly have that $0 <$

$s_n - \ln(n+1) < 1$, and hence

$$\ln(n+1) < s_n < \ln(n+1) + 1 \tag{11.1}$$

holds (explain this reasoning carefully). Also, since $s_n - \ln(n+1)$ increases and is bounded above by 1, *it does have a limit less than or equal to 1*, which we will study in the next section. [A theorem from analysis guarantees that such a sequence (bounded above and increasing) does indeed converge, and to a number not exceeding the upper bound.]

This now gives us a bound depending on n for the partial sums, s_n. And inequality (11.1) also suggests that a reasonable first approximation for $J(n)$ would be the smallest integer greater than or equal to $\ln(n+1)+1$. certainly this would be an upper bound for s_n and it is *not very far above.*

Exercise 14: Use this estimate to redo Exercise 6.

Exercise 15: Let's see if we can improve things even further! Consider the right-hand diagram in Figure 11.6. In this picture, we have replaced the slightly curved edges of the "triangular" regions of the left-hand picture with straight lines, so that we now have actual triangles (whose exact area we can find). What is the exact area which the total of the shaded rectangles approaches as n approaches infinity? Carefully explain your reasoning.
[Hint: You might want to look back at Exercise 3.]

From the way that the regions differ in the two diagrams of Figure 11.6, it is clear that the shaded area, $s_n - \ln(n+1)$, is larger (though only slightly) in the left-hand diagram than in the right. Hence, our limit is greater than $1/2$ and well less than 1.

Exercise 16: In Table 3, fill in the column s_n with the values from your work in Table 2. Then with a calculator, fill in the logarithm column and the difference column. In the next section, we will further explore this sequence using the computer. It will lead us to discover an important constant in mathematics.

n	s_n	$\ln(n+1)$	$s_n - \ln(n+1)$
3			
10			
100			
500			
1000			

Table 3

11.5 Euler's constant

We want to find a better estimate for the number $J(N)$ from Section 11.3. (Recall that N is the number of terms that will be computed by tomorrow at this same time.) To do this, we go again to the computer. We need a program to calculate the difference $s_n - \ln(n+1)$ more precisely than we have done so far. We call the program EULERC, for reasons that will become clear shortly.

Program outline: EulerC

```
Input: A positive integer N.
Output: Prints a table with ROWS:
        n, SUM(n), ln(n+1), DIFF=SUM(n)-ln(n+1).
        (1<=n<=N)

RUN Sequences and Series to get sk
FOR n = 1 TO N
  PRINT n.
  PRINT sk(n).
  PRINT ln(n+1).
  PRINT the difference: DIFF = SUM(n)-ln(n+1).
END FOR
```

Question 1: Use EULERC to check your values in Table 3 for DIFF = $s_n - \ln(n+1)$ for $n = 50$ and 100. What pattern do you notice for the values of DIFF as n gets large? Run EULERC again, this time letting ROWS =

500. Check your values in Table 3 for DIFF $= s_n - \ln(n+1)$ for $n = 500$ and $n = 1000$. What is the value of DIFF for $n = 2000$?

The number to which these DIFF values converge is a famous one, called **Euler's constant**. It is the number, that we'll denote C, that is defined as follows.

$$C = \lim_{n\to\infty} (s_n - \ln(n+1)).$$

The convergence is very slow, and we know that $C > 1/2$. (Remember our geometric suggestion that the shaded area DIFF of the square U was slightly greater than 1/2).

Question 2: Use EULERC to answer the following two questions:

1. At what n does the value of DIFF become greater than 0.5? Since we showed that the values of DIFF form an increasing sequence, you know that DIFF is then greater than 0.5 for all succeeding n's. That is, for all values of n beyond the one you found, $0.5 < s_n - \ln(n+1) < 1$.

2. Similarly, for what value of n is DIFF > 0.54? DIFF > 0.55? DIFF > 0.56? DIFF > 0.57? DIFF > 0.574? ..., etc. Make a chart showing these values as in Table 4. Note that the last value in the table (where DIFF > 0.5772) will require a fairly large amount of computing—its value is 31,918.

DIFF	n
0.500	
0.550	
0.560	
0.570	
0.574	
0.575	
0.576	
0.577	
0.5771	
0.5772	

Table 4

The actual value of Euler's constant is known to be $C = 0.57721567...$ It is still not known if this constant is an irrational number!

Now, we know that $s_n - \ln(n+1) \to C$ and this converging sequence is *increasing*, so we have that for all n, $0 < s_n - \ln(n+1) < C$. Hence

$$\ln(n+1) < s_n < \ln(n+1) + C \tag{11.2}$$

gives us the best upper bound that we could expect for the numbers s_n.

Question 3: So now you can answer the central question of this chapter with *much* better results. Answer the following questions.

(a) What does inequality (11.2) suggest as the way to calculate $J(n)$?

(b) From your earlier calculation in Exercise 12, what is the number N of terms that will be computed by this time tomorrow? What is $J(N)$?

(c) What interval (from 0 to an integer upper bound $J(n)$) would you use if you were going to plot the harmonic series to 1 million terms? To 1 billion terms? That is, what are $J(1,000,000)$? $J(1,000,000,000)$?

(d) This is a calculator problem (too large for the computer!). Use the facts you have developed to estimate the number of terms in the harmonic series whose partial sums are contained in the intervals:

$$[1,4], [1,5], [1,6], [1,7], [1,8], [1,9], [1,12], [1,16], [1,20].$$

[Hint: Use the fact that e^x is the inverse of $\ln(x)$.]

(e) Use the program SEQUENCE AND SERIES to test your results in the previous question (where appropriate—What do we mean by "where appropriate"?). Where deemed appropriate, what are the actual values of n and s_n for which the partial sums fail to fall in the given intervals. Show a careful chart of these values.

11.6 Additional exercises and questions

Use a computer (SEQUENCE AND SERIES is ideally suited for this) to explore some of the following series. See if you can guess whether the limit of the series exists, and (if so), what that limit is. See if you can prove your observations.

1. $\displaystyle\sum_{k=1}^{\infty} \frac{(-1)^{k+1}}{k} = 1 - \frac{1}{2} + \frac{1}{3} - \frac{1}{4} + \frac{1}{5} - \ldots$

2. $\displaystyle 4\sum_{k=1}^{\infty} \frac{(-1)^{k+1}}{2k-1} = 4 - \frac{4}{3} + \frac{4}{5} - \frac{4}{7} + \ldots$

3. $\displaystyle 6 + \frac{6}{4} + \frac{6}{9} + \frac{6}{16} + \frac{6}{25} + \ldots$
   [Hint: This limit is a power of π.]

4. $\displaystyle 90\sum_{k=1}^{\infty} \frac{1}{k^4}$
   [Hint: And another power of π.]

5. $\displaystyle 12\sum_{k=1}^{\infty} \frac{(-1)^{k+1}}{k^2}$
   [Does this answer relate to an earlier answer?]

6. $\displaystyle\sum_{k=1}^{\infty} \frac{1}{k(k+2)}$
   [See if you can prove this result. It's similar to the example in Exercise 3.]

7. $\displaystyle\frac{1}{1\cdot 3} + \frac{1}{3\cdot 5} + \frac{1}{5\cdot 7} + \frac{1}{7\cdot 9} + \ldots$
   [Again, prove this result in a similar fashion.]

8. $\displaystyle\sum_{k=1}^{\infty} (-1)^k \frac{\sin(k)}{k}$

9. $\displaystyle 1 + 1 + \frac{1}{2!} + \frac{1}{3!} + \frac{1}{4!} + \ldots$
   [Note: The factorial function is 1 at 0 (i.e., $0! = 1$).]

10. $\displaystyle\sum_{k=0}^{\infty} \frac{(-1)^k}{k!}$
    [How does this relate to the previous example?]

Question 4: In Exercise 11, you were asked to speculate on the behavior of the ratios of the successive entries in the first column of Table 2. Can you support your speculations with mathematical analysis? This is a somewhat more challenging question than others we have looked at.
[Hint: Observe that the entries in the first column are those integers n for which $J(n)$ takes the values $1, 2, 3, \ldots$. We have seen that for large n

$$J(n) \approx \ln(n+1) + C$$

where C is Euler's constant. Use this expression to write n in terms of k if $J(n) = k$ and k is large. Similarly, express m in terms of k if $J(m) = k + 1$. What can you say about the limiting value of the ratio m/n as k approaches infinity?]

11.7 Computer programs

11.7.1 *MATLAB/Octave* programs

Program: Sequence and Series

```
function [ak,sk]=seqser(f,n,ploton)
%function [ak,sk]=seqser(f,n,ploton)
%
%Computes sequence of terms and partial sums for a series and
%plots the result
%
%Inputs
%  f - a function f that gives the terms (f(k)=a_k)
%  n - number of terms to use
%  ploton - option input: if 1 then plots, if 0 no plot produced
%           plot produced by default
%Output
%  ak - a vector of terms of the series
```

```
%  sk - a vector of partial sums

if nargin < 3
   ploton=1; %if ploton input not used plot
end; %if

ak=zeros(1,n); %initialize vector of terms
pk=zeros(1,n); %initialize vector of partial sums
ak(1)=f(1);    %first term
sk(1)=ak(1);   %first partial sum
for k = 2:n
   ak(k)=f(k);
   sk(k)=sk(k-1)+ak(k);
end; %for

%generate plots
if ploton
   ks=1:n;
   subplot(2,1,1)
   plot(ks,ak,'k.')
   ylabel('a_k')
   xlabel('k')
   set(gca,'FontSize',20)

   subplot(2,1,2)
   plot(ks,sk,'k.')
   ylabel('s_k')
   xlabel('k')
   set(gca,'FontSize',20)
end; %if
```

This program requires the input of the function to give the terms of the sequence. The easiest way to do this is to define an anonymous function. For example, if we want to compute and plot 100 terms of the harmonic series enter:

```
>> f = @(x) 1/x
```

```
>> [ak,sk]=seqser(f,100,1)
```

Program: Stopwatch

```
%Script file
%the number of terms of the harmonic series summed by
%the computer after each six-second interval up to one
%minute

sk=0;  %the partial sum so far
k=0;
tic; %start timer
for i=1:10;
  while toc<6*i
    k=k+1;
    sk=sk+1/k;
  end; %while
  fprintf('Time used is: %6.2f sec. \nNumber of terms summed: %d \n\n',toc,k)
end; %for
```

Program: EulerC

```
function []=eulerc(N)
%function []=eulerc(N)
%
%Produces a table of outputs for the harmonic series for
%A row for each value of n (1<=n<=N) contains the value
%of n, s_n (the nth partial sum), ln(n+1), and s_n-ln(n+1)
%
%Inputs
%  N - number of rows
%Output
```

```
%  none, a table is printed

f = @(x) 1/x; %function for term

[ak,sk]=seqser(f,N,0); %get sequence of partial sums
k=1:N; %vector of k values
lnk=log(k+1); %vector of ln of k+1 values
diffk=sk-lnk; %difference

%loop through and print results
for n=1:N
   if mod(n,20)==1 %every 20 terms print column labels
      fprintf('\n n \t \t  s_n \t \t ln(n+1) \t s_n-ln(n+1) \n')
   end; %if
   fprintf('%6g  \t %9g  \t  %9g  \t %12.12g \n',k(n),sk(n),lnk(n),diffk(n))
end
```

11.7.2 *Mathematica* programs

Sequences and Series

The first input is the sequence a_k you are summing. You must use the variable name k. The sums are computed starting at the second input and ending at the third input.

```
SeqSer[expr___, start_, end_]:=Module[{st = start, gr1, gr2, final = end, n, s, fct = expr},
s[st − 1] = 0; s[n_]:=s[n] = N[s[n − 1] + fct/. {k → n}]; gr1 = Print@Graphics[
Table[{Blue, AbsolutePointSize[5], Point[{n, s[n]}]},
{n, st, final}], Axes → {Automatic}, AxesOrigin->{0, 0}, PlotRange → All,
PlotLabel → Text ["s_n"] , BaseStyle → {FontSize → 14}, AspectRatio → .3,
ImageSize → 600]; gr2 = Print@Graphics[Table[{Red, AbsolutePointSize[5],
```

Point[{n, fct/. {k → n}}]}, {n, st, final}], Axes → {Automatic},

AxesOrigin->{0, 0}, PlotRange → All,

PlotLabel → Text ["a_n"],

BaseStyle → {FontSize → 14}, AspectRatio → .3, ImageSize → 600]]

SeqSer[1/k, 1, 20]

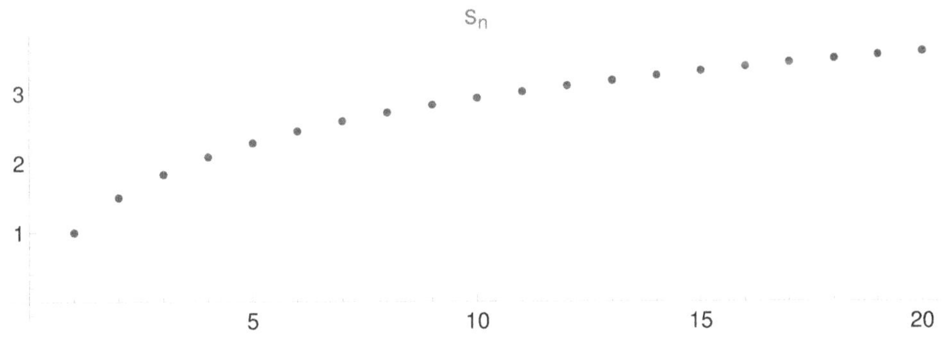

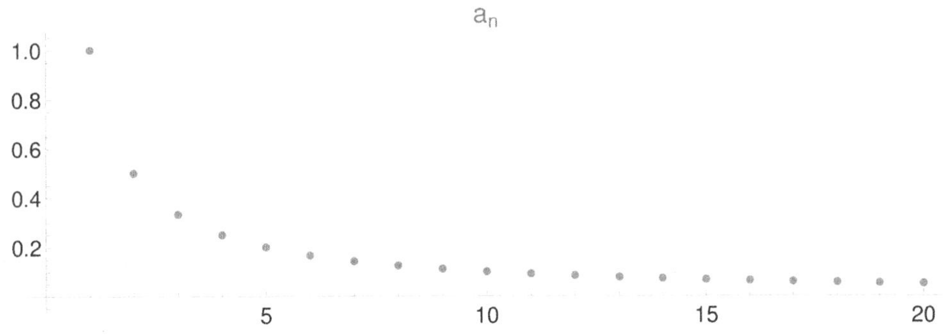

If you want to change the code above (and below) to work for the sequence $a_k = 12\frac{(-1)^{k+1}}{2k-1}$ for example, type "12 (-1)^(k+1)/(2k-1)" instead of "1/k".

Partial Sums and J(n)

Next we compute (numerical) partial sums of the harmonic series and $J(n)$

NSum[1/k, {k, 1, 10}]

2.92897

NSum[1/*k*, {*k*, 1, 11}]

3.01988

***J*[n_]:=Ceiling[NSum[1/*k*, {*k*, 1, *n*}]]**

***J*[11]**

4

Stopwatch

The command below runs for 60 seconds of CPU time.

Every 6 seconds it tells you how many terms of the partial sum of the harmonic series have been computed so far, and what the partial sum is at this point of the computation.

′time′ is the time interval between printouts; 'trials' is the number of printouts.

```
s = 0.; k = 1; time = 6; trials = 10;
Do[T = TimeUsed[];
While[TimeUsed[] − T < time, s = s + 1/k; k++]
Print["Time used: ", time ∗ i, " sec.\nNumber of terms summed: ",
NumberForm[k, DigitBlock → 3], "\nSum: ", s], {i, 1, trials}]
Clear[s, k];
```

Euler Constant

"EulerC[n]" computes the quantity $\sum_{k=1}^{n} \frac{1}{k} - \ln(n+1)$

```
EulerC[n_]:=N[Sum[1/k, {k, 1, n}] - Log[n + 1], 20]
```

```
EulerC[310]
```

0.57560708627592750822

Chapter 12

Experiments in Periodicity

12.1 Introduction

Periodic functions are all around us. They describe phenomena that repeat over and over again, like the rising of the sun, the oscillation of a spring, and the motion of sound. The trigonometric functions are the most familiar periodic functions. They can be combined to make new functions. We can also produce new functions by differentiating and antidifferentiating. In this chapter, we will use the computer program *Desmos* to explore properties of area accumulation functions (antiderivatives) of periodic functions.

We begin by establishing some language to describe the functions we will study. A function $y = f(x)$ is said to be p-**periodic**, provided there exists a positive real number p such that for every real number x,

$$f(x + p) = f(x)$$

The smallest positive p such that f is p-periodic is called the *period* of f. For example, the sine and cosine functions are 2π-periodic. We will do some experiments with sines and cosines and with some functions which are derived from them.

Exercise 1: What is the period of each of the six trigonometric functions?

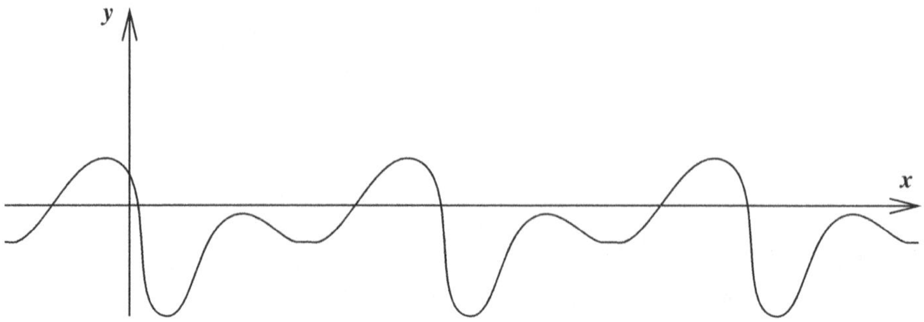

Fig. 12.1: A periodic function

Question 1: Using the trigonometric functions as building blocks, make new functions: by adding, multiplying, composing, etc. [An intriguing one to try is $\sin(x+\cos(x))$.] Use a graphing utility (for example *Desmos'* graphing calculator) to check your new functions for periodicity.

Question 2: You can also make new functions from old by differentiating and antidifferentiating. Try these calculus operations on some of your examples of periodic functions from Question 1. Can you predict when a new function will again be periodic? If the new function is periodic, can you predict its period?

As you start thinking about trying to support some of your findings in Questions 1 and 2, you should think geometrically, as well as algebraically. Specifically, when you think about adding functions, or multiplying by constants, think of adding y-coordinates or about scaling x or y. Similarly, when you think about the derivative, think about slope as well as about differentiation.

Exercise 2: Show that the derivative of a periodic function is periodic, with the same period. Give two arguments, a formal one using differentiation and an informal one using slopes.

The antiderivative has a geometric representation too, and we're going to make heavy use of it. By the fundamental theorem of calculus, a continuous function f always *has* an antiderivative, namely one given by an area accumulation function.

> For any real number a, we define an **area accumulation function** A for f whose value $A(x)$ is the signed area between the graph of f and the x-axis over the interval from a to x.

Here *signed* means that regions below the x-axis contribute a negative amount, above give a positive contribution. (We can even give meaning to $A(x)$ for $x < a$ by taking the negative of the signed area between the graph and the x-axis from x to a.)

We will use the program *Desmos* to investigate area accumulation functions. The next section will walk you through the software for some particular functions, but now we lay out some of the questions you will be exploring using the program. At this point, you should think the questions through *without* using the computer. Later, you should return to the questions and use the computer to try many examples.

As you probably recall from calculus, a function can have many antiderivatives. In fact, you can produce many different area accumulation functions by varying your choice of the starting point a.

Question 3: Examine the area accumulation function for some of your examples of periodic functions. Observe the effect of varying the starting point a. If your function $f(x)$ is positive for x's near a, what is the effect on $A(x)$ of making a larger? smaller? What about if $f(x)$ is negative near a? If you start with a periodic function, is the area accumulation function periodic? Can you make it periodic by adjusting the choice of a?

Obviously changing the function $f(x)$ will change the area accumulation function too.

Question 4: Suppose you replace $f(x)$ by $f(x) + M$ for some constant M. If $f(x)$ is periodic, is $f(x) + M$ periodic? How does the value of M affect the area accumulation function?

12.2 Area accumulation

In this section we use *Desmos* to look at area accumulation functions and try to determine periodicity.

- Open the *Desmos* graphing calculator:

  https://www.desmos.com/calculator.

- Enter your function using function notation. For example, if we are interested in the area accumulation for $\sin(x)$ enter:

  $$f(x) = \sin(x)$$

  This will create a graph of the function.

- Next we create an inequality to shade the area. On the next line enter:

  $$\{f(x) > 0 : 0, f(x)\} \leq y \leq \{f(x) > 0 : f(x), : 0\}\{-8 \leq x \leq b\}$$

  To get the symbol $\leq$ in *Desmos*, type $<$ followed by $=$. Similarly, to get $\geq$ enter $>$ followed by $=$. You will be prompted to create a slider for b, do it. The above code uses *Desmos'* if-else structure. Specifically, a statement of the form

  $$\{cond : expression1, expression2\}$$

  returns $expression1$ when $cond$ is true and $expression2$ when $cond$ is false. Thus our statement becomes: $0 \leq y \leq f(x)$ $\{-8 \leq x \leq b\}$ when $f(x)$ is positive and $f(x) \leq y \leq 0$ $\{-8 \leq x \leq b\}$ when $f(x)$ is not positive.

- Create the graph of the area accumulation function by entering:

  $$A(x) = \int_{-8}^{x} f(t)dt$$

  To get the integral symbol, type: *int*. *Desmos* uses numerical integration to create the graph, similar to the methods covered in Chapter 10.

- Finally, we will add a point to our accumulation function that indicates the shaded area by entering:

  $$(b, A(b))$$

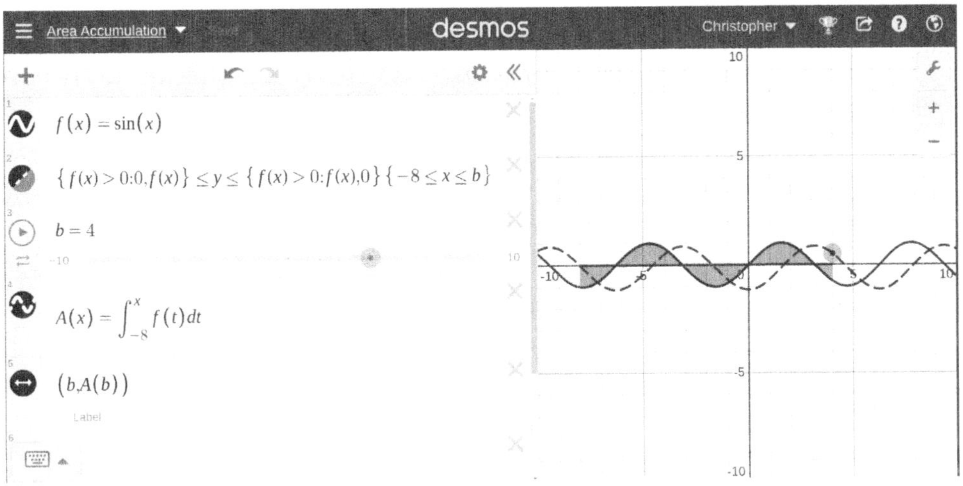

Fig. 12.2: Desmos showing area accumulation for $y = \sin(x)$ *over* $[-8, 4]$

- The result is shown in Figure 12.2 (the colors and line styles have been changed using the gear icon). You can move the slider to see the relationship between the area and the accumulation function $A(x)$.

Exercise 3: Now we consider the possible periodicity of $A(x)$. Use *Desmos* to fill in the chart (Table 1). You can get values of A by entering, for example, $A(0)$. To get π, type *pi*. Because *Desmos* uses numerical integration to evaluate A, the results may not be exact. Do the results in Table 1 and Figure 12.2 suggest periodicity for the function $A(x)$?

x	$A(x)$
-2π	
$-\pi$	
0	
π	
2π	
3π	

Table 1: Accumulating sums

Exercise 4: You know that $-\cos(x)$ is an anti-derivative for $\sin(x)$. Does the graph of $A(x)$ that you see in *Desmos* look exactly like this antiderivative? Explain your answer.

We want to modify our function so that the resulting area accumulation function is precisely the function $-\cos(x)$. We are using

$$A(x) = \int_a^x \sin(t)\,dt$$

One way to change $A(x)$ is by changing the left-endpoint a. We want to begin modestly by changing a only very slightly, so that we begin with the graph of $f(x) = \sin(x)$ over an interval very nearly the same as the previous $[-8, 12]$.

Exercise 5: Modify our *Desmos* function for $A(x)$ to be an integral starting at a rather than -8. *Desmos* will prompt you to add a slider for a. Experiment with the left hand endpoint, a, of the domain of f so that it is close to, but not equal to, -8. Observe how the graph of the function $A(x)$ differs, if indeed it does, from what you sketched and plotted previously. For instance, try $a = -8.25$, $a = -8.5$, $a = -7.75$, $a = -7.5$, $a = -2\pi$ (enter "$-2pi$" precisely that way). Write a paragraph describing how the graph of the area accumulation function $A(x)$ varies as you change the left endpoint a. Discuss *why* this phenomenon occurs.

Exercise 6: Still using the same function, find a point a that comes as close as possible to the place for which the area accumulation function $A(x)$ equals $-\cos(x)$. Can you prove that your empirical choice is correct?

Compare your results in Exercises 5 and 6 to your analysis in Question 3.

Question 5: This is a non-computer question. Please do it *before* you move on to the next exercise. Suppose you were to alter the function, $y = f(x)$, by adding a constant to it. For instance, suppose you looked at $f(x) = \sin(x) + 0.1$ and sought its area accumulation function $A(x)$ from some initial point, for instance $a = -8$. How do you think this would change the appearance of the graph of $A(x)$? Think about this carefully. Sketch some pictures:

first using $\sin(x)+0.1$ and then considering the similar function $\sin(x)-0.1$. Carefully justify your assertions here.

Exercise 7: Now continue using *Desmos*, reset a to -8, and plot $A(x)$ using first $f(x) = \sin(x)+0.1$ and then $f(x) = \sin(x)-0.1$. Repeat this for several other additive constants. Verify your answer to Question 5. Is $A(x)$ periodic? Explain your answer.

Look again at your thinking in Question 4.

12.3 A new type of function

Enter the function $f(x) = \sin(x + \cos(x))$ into the *Desmos* graphing calculator. Be sure to type this exactly in this way, taking $x + \cos(x)$ as the *argument* of the sine function.

Exercise 8: Plot this function over the region $[-8, 12]$. Does this function appear to be periodic? If so, what is your guess for the value of the period p?

Exercise 8 here begs an answer of "yes." Certainly, a function of the form $\sin(x + K)$ is 2π-periodic where K is a constant. (Why?) Looking at the function under consideration, however, you might think that it should not be periodic, since it is the sine of the quantity *x plus something that is not constant.* To study this question further, notice that f is a composition: the sine function applied to $g(x) = x + \cos(x)$. It would be interesting to plot the function $g(x)$.

Question 6: Describe the graph you get. Is this function periodic? Justify your answer in terms of the *shape* of the curve g and the heights at which g "levels out." Speculate on what might be at work here. What terminology might you use to describe the function g? Keep the graph on the screen for the next question.

Question 7: On the same axes graph g from Question 6 with the function $f(x) = \sin(x + \cos(x))$. Write a careful description of how the *shape* of the graph of f is affected by the *shape* of the graph of g.

Exercise 9: Use the definition of a periodic function to *prove* that $f(x) = \sin(x + \cos(x))$ is 2π periodic.

If a function is p-periodic, then $f(x+p) - f(x)$ is always zero. We might view a function as nearly periodic if the difference is non-zero but constant. This motivates the following definition. Suppose $p > 0$. We say a function $f(x)$ is **linearly p-periodic** if and only if there exists a real constant M such that for all x,

$$f(x+p) = f(x) + M$$

The constant M is called the **p-translation constant** for f.

Exercise 10: Sketch the graph of a continuous linearly p-periodic function f for which $p = 1$ and the p-translation constant M is 0.5.

With this definition in mind, look again at Question 6.

Exercise 11: Suppose $y = F(x)$ is continuous and linearly p-periodic. Prove that its derivative, $y = F'(x)$, is p-periodic.

12.4 Antiderivatives of periodic functions

In this section we will attempt to characterize those functions that can be antiderivatives of periodic functions. We begin with a computer example.

Exercise 12: Using *Desmos*, plot the function $f(x) = \sin(x+\cos(x))$, along with its area accumulation function

$$A(x) = \int_{-8}^{x} f(t)dt$$

1. Notice that while there are occasional places where this $A(x)$ decreases, it generally has an upward trend to it. Give a short, but careful, explanation as to *why* this happens.

2. We saw in Exercise 7 that a way to change that "upward trend" to a "downward trend" or even a fairly "level trend" in the graph of $A(x)$ was to modify the function f by adding or subtracting a constant. Experiment with this function to produce an $A(x)$ that *appears* to be periodic. What is the constant, call it Q, that works here? That is, what value of Q makes the area accumulation function of $\sin(x + \cos(x)) + Q$ periodic?

3. This part will be for use in a later section. Use *Desmos* to evaluate the integral:

$$\int_0^{2\pi} f(x)\,dx.$$

You can do this by entering the expression on its own line. The function f needs to be defined on another line (Recall that you get the integral symbol by typing *int* and π by typing *pi*.) What value do you get?

4. Try integrating this function over a few other intervals of length 2π. What do you notice?

Exercise 13: Plot the function $f(x) = \sin(x + \cos(x)) + Q$, along with its area accumulation function

$$A(x) = \int_a^x f(t)dt$$

Start with $a = -8$.

1. If you change the starting point of the graph of f from $a = -8$ to any other number, will that change the fact that its antiderivative $A(x)$ is periodic? Why or why not? Use the computer to experiment here.

2. Use the computer to find a starting point a such that $A(x)$ is *never negative*. What is such a point? Within any interval of length p (the period of f) is there just one such point, or many? As you do this, think carefully about *why* you pick the point(s) that you choose.

3. Similarly, find a starting point at which $A(x)$ is *never positive*. Consider the same issues here that you did in the previous part. Why do these starting points work in this way?

Now revert to the left endpoint of $a = -8$.

Your earlier work (Exercise 12) suggested that a p-periodic function whose area accumulation function $A(x)$ was not periodic could be altered by adding a constant Q so that the resulting area function was p-periodic. In notation, if $y = f(x)$ is p-periodic with area function $A(x)$, then there exists a constant Q such that the (still) p-periodic function $f_1(x) = f(x)+Q$ has a p-periodic area accumulation function, $A_1(x)$. While this is "suggested" by your computer work, the findings of the next section put this on solid ground.

12.5 Finding the periodic antiderivative

Exercise 14: Now we look at another function over the interval $[-8, 12]$ with its area accumulation function. Change f to

$$f(x) = \sin(x + \cos(x + \sin(x)))$$

and plot this function. The picture gives a somewhat tooth-like appearance.

1. *Prove* that $f(x)$ is periodic. What is its period p?

2. As you did in Exercise 12, experiment with the computer to find a real constant Q such that $\sin(x + \cos(x + \sin(x))) + Q$ has a periodic area accumulation function. What value do you get for Q?

3. Letting p be the period you found for the function f, use *Desmos* to calculate

$$\int_0^p \sin(x + \cos(x + \sin(x)))\, dx,$$

and keep this value for future use.

4. Try integrating the same function over a couple of different intervals of length p. What do you observe?

Exercise 15: Repeat Exercise 14 for the following functions, and fill in the values from these and the previous two exercises in Table 2.

1. $\sin(x + \sin(x + \sin(x)))$
2. $\cos(x + \sin(x + \cos(2x)))$
3. $\sin(2x + \cos(x + \sin(x)))$

Try to find other functions of this type that you feel might be periodic.

$f(x)$	Q	Period	Integral over a period
$\sin(x + \cos(x))$			
$\sin(x + \cos(x + \sin(x)))$			
$\sin(x + \sin(x + \sin(x)))$			
$\cos(x + \sin(x + \cos(2x)))$			
$\sin(2x + \cos(x + \sin(x)))$			

Table 2: Summary of functions

Question 8: Examine the data is Table 2 and see if you can determine the relationship, if any, between the constant Q that you add to f to make its antiderivative p-periodic and the integral over one period of the function f.

Now the goal is mathematical analysis that will put your computer conjectures on a solid foundation.

Question 9: If f is a p-periodic function, what can you say about the value of

$$\int_x^{x+p} f(t)\,dt$$

as x varies? Can you *prove* that what you observed in specific examples holds in all cases?

Question 10: Suppose f is a p-periodic function. To keep everything nice, assume that f is defined and continuous for all real numbers. Fix a choice of a, and let

$$A_1(x) = \int_a^x f(t) + Q\, dt$$

Can you *prove* that it's always possible to choose Q so that $A_1(x)$ is p-periodic? [Hint: Make use of the answer to Question 9.]

Exercise 16: Using the theorem that is implied in Question 10, check the results of your work in Exercises 12, 14 and 15 (as given in Table 2).

12.6 Further investigation

There are many directions that your further study could take—both computer experiments and pencil and paper investigations. You can, for instance, characterize p-periodic functions in terms of properties of an antiderivative, as in the following theorem.

Theorem 1: A continuous function $y = f(x)$ is p-periodic if and only if any antiderivative R of f is linearly p-periodic.

The functions we have been looking at in the last sections of this chapter have been made up of *compositions* of 2π-periodic functions, for instance $f(u) = \sin(u)$, with linearly 2π-periodic functions, as $u = g(x) = x + \cos(x)$, and we have seen the result to be 2π-periodic. We can make a general statement of this in Theorem 2.

Theorem 2: If a continuous function $y = f(u)$ is p-periodic in u, and if $u = g(x)$ is linearly p-periodic with p-translation constant kp, where k is any integer, then $y = f(g(x))$ is p-periodic.

See if you can prove these theorems.

Further interesting directions involve speculation as to the nature of p-periodic functions made up, as we have in this chapter, of sines and cosines linked together, such as:

$$y = \sin(x + \cos(x + \sin(x + \cos(x + \cos(x)))))$$

Any combination of sines and cosines of this nature gives a different periodic function (thus generating an *infinite* class of periodic functions). There are many fascinating characteristics of such functions, and you may have fun studying them.

Chapter 13

Iteration to Solve Equations

13.1 Introduction

In Chapter 1 we saw that when a sequence of iterates

$$x_0, f(x_0), f(f(x_0)), \ldots$$

converges to a limit L, then L is a solution of the equation

$$f(x) = x.$$

This suggests that we might try solving an equation $g(x) = 0$ by rewriting it in the form

$$g(x) + x = x$$

and performing iteration on the function f defined by $f(x) = g(x) + x$. Any limit produced in the process, for a specific initial number x_0, yields a solution of the equation $g(x) = 0$.

We first want to consider how our experience from Chapter 1 carries over when the function g is linear, say $g(x) = mx + b$. The function f which gets iterated is then also linear and has the form

$$f(x) = (m + 1)x + b.$$

In this situation there are just two possibilities: either the iterates $x_0, f(x_0)$, $f(f(x_0)), \ldots$ converge for every choice of x_0 or they converge only when x_0 happens to be a solution of $g(x) = 0$.

For nonlinear equations the situation is somewhat more complex, since nonlinear equations may have more than one solution. Quadratic equations may have as many as two, cubic equations as many as three, and the equation $\sin(x) = 0$ has infinitely many solutions. Thus there is a strong possibility that the limit of a convergent iterative process, and therefore the particular solution which we obtain, may be dependent on the particular initial guess x_0.

Here is a description in pseudocode of a program which uses iteration to solve $f(x) = x$.

Program outline: Solver

```
Input:   a function f(x), the number n of iterates,
         and an initial value x0
Output: the n iterates of f (and thereby a solution
         of f(x) = x)

INITIALIZE its as an (n+1)-dimensional vector
its(1) := x0            %set first entry of its to x0
FOR i = 1 TO n
    its(i+1) := f(its(i))
END FOR
OUTPUT its
```

It will be obvious from the graph of $y = f(x)$ that the equation $f(x) = x$ has exactly three solutions. For the initial value $x_0 = 0.5$ chosen, iteration of $f(x)$ apparently gives convergence to one of these. You should try other choices for x_0 in order to attempt to get convergence to the other solutions. Are you successful?

In studying nonlinear functions under iteration, you will find that some things are similar to the linear case and you will also note some significant differences: the trick is to clearly sort out the similarities and the differences. In Chapter 14 we saw that a nonlinear (quadratic) f could lead to chaotic behavior under iteration, but we will concentrate our attention here on obtaining convergent behavior when we begin the iterative process close to a fixed point. We will attempt to extrapolate from our experience with linear functions in Chapter 1. A key observation for thestudy of nonlinearity is that *any* differentiable function looks *locally* like a linear function. This is more or less the definition ofdifferentiability: tangent lines are local linear approximations. Our experience in Chapter 1 should suggest that the slope of f at a fixed point has a lot to do with convergence to that point. The following exercises are designed to test this conjecture.

Exercise 1: Try $f(x) = \cos(x)$. Sketch the graphs of $y = x$, $y = \cos(x)$. Do they meet? If they do, a value of x for which this happens is asolution of $\cos(x) = x$. Choose some initial values x_0 and iterate f. Observe!

Exercise 2: What about $f(x) = \sin(x)$? Sketch the graphs $y = x$, $y = \sin(x)$; what solution are you looking for? What happens if you iterate here? What if $f(x) = \sin(2x)$? Try to explain what happens from your experience with the iteration of linear functions.

Exercise 3: Find as many solutions of the equation $x^3 - 2x + 1 = 0$ as you can by iteration. Try to explain the success or failure of the iterative process for each of the solutions.

Exercise 4: If a is a positive constant (e.g., $a = 2$) you can try solving the equation $g(x) = x^2 - a = 0$ by iterating the function $f(x) = x^2 + x - a$. Try this out. How well does it work?

Exercise 5: Try to solve the same equation by instead iterating $f(x) = a/x$. After all, if $a/x = x$ then ... How does the iteration work out for this function?

Exercise 6: Show that the fixed points of the function $f(x) = (x + a/x)/2$ are the solutions of the equation $x^2 - a = 0$. How well does iteration work for f in locating the fixed points? (You will learn in the next section how this function was obtained.)

Exercise 7: Try $f(x) = 2\cos(x)$. Sketch graphs, and locate a solution of $2\cos(x) = x$. What happens when you iterate?

Exercise 8: Did you get a mess? Well, if you choose $f(x) = (x+2\cos(x))/2$ you are still solving $2 \cdot \cos(x) = x$. But is this choice more successful?

13.2 Improving convergence

We have seen in the above exercises that the value of the derivative at a fixed point very much determines the convergence of the iterative process, just as it it does in the linear case. But if convergence fails, are we without recourse in calculating precise values for the fixed points?

It turns out that a very simple device will give success in most instances. Instead of iterating f, we iterate the function

$$h(x) = \lambda f(x) + (1 - \lambda)x, \quad \lambda \neq 0,$$

where λ is a number which we choose at our discretion. This gives a weighted average (where the weights λ and $1 - \lambda$ sum to 1) of the function f and the *identity function* $\mathrm{id}(x) = x$.

Exercise 9: Show that the fixed points of h are exactly the same as the fixed points of f.

Exercise 10: Find a choice for λ so that iteration will yield convergence to the fixed point of $f(x) = 2x - 1$ no matter which initial value is chosen. Show that there is an optimal choice of λ, one which gives the most rapid convergence.

Exercise 11: For a nonlinear function with several fixed points, good choices for λ will vary from fixed point to fixed point. Return to the situation of Exercise 3 and try to find good values for λ for each of the points where convergence failed before.

We can even automate the choice process for λ. If c is a fixed point of the function h, most rapid convergence occurs if $h'(c) = 0$ (at least, this is true when h is linear). This suggests choosing λ so that this is approximately true, i.e., if x_{i-1} is an approximation to a fixed point, we obtain the choice λ so as to make $h'(x_{i-1}) = 0$ and then compute the next iterate $x_i = h(x_{i-1})$. This means that we are changing the choice of λ as we proceed in the process, which is a new feature, but one which doesn't really complicate things. In terms of f this process requires that

$$\lambda f'(x_{i-1}) + (1 - \lambda) = 0,$$

which implies

$$\lambda = \frac{1}{1 - f'(x_{i-1})},$$

and

$$x_i = h(x_{i-1}) = \frac{f(x_{i-1}) - x_{i-1} \cdot f'(x_{i-1})}{1 - f'(x_{i-1})}.$$

Exercise 12: Apply the above scheme to the solution of the equation $g(x) = x^2 - 3 = 0$, i.e., to the situation where $f(x) = x^2 + x - 3$. How successful is it in providing convergence to the solutions of the equation? How does the solution obtained depend on the initial choice x_0?

13.3 Questions to explore

In answering the following questions, you may use some of the examples provided in the exercises and, even better, use additional examples of your own invention to demonstrate what you believe is true.

Question 1: Under iteration, what are the most significant similarities and differences between the linear and nonlinear cases?

Question 2: Under what conditions does the iteration of a function f result in convergence to a fixed point provided that our starting point is sufficiently close to that fixed point?

Question 3: Under what conditions does interation fail to converge to a specific fixed point no matter how close our starting point is to that fixed point?

Question 4: In relation to the preceding two questions, what are the borderline cases? Can you find new phenomena here which were not present in the iteration of linear functions?

Question 5: Under what circumstances can the iteration process be modified to ensure convergence to a given fixed point?

Question 6: What can you say about the rate of convergence to fixed points? Can you give examples where the convergence is extremely fast or extremely slow? How does the convergence rate for modified iteration compare with the convergence rate for simple iteration?

13.4 Computer program

13.4.1 *MATLAB/Octave* program

Solver

```
function its=solver(f,x0,n)
%function its=solver(f,x0,n)
%
%Estimates solution to f(x)=x using iteration
%
%Inputs
%  f - function
%  x0 - left endpoint of integral
%  n - number of iterates
%Output
%  its - vector of iterates

its=zeros(1,n+1);
its(1)=x0;
for i=1:n
   its(i+1)=f(its(i));
end; %for
```

This program requires the input of the function to be solved. The easiest way to do this is to define an anonymous function. For example, if we want to estimate the solution $2\sin(x) = x$ with $x_0 = 0.5$ and $n = 10$ iterates, enter:

```
>> f = @(x) sqrt(x^2 +1)
>> its=solver(f,0.5,10)
```

Chapter 14

Iteration of Quadratic Functions

14.1 Introduction

In Chapter 1 you iterated a linear function and found out that its behavior was pretty simple. In this chapter you will iterate quadratic functions and find that their behavior is not at all simple! What you learned by iterating linear functions does cast some light on the non-linear case, but you will also encounter some completely new phenomena. In fact, iteration of quadratic functions can lead to the the erratic behavior that mathematicians call *chaos*.

14.2 Some theory

Before you begin your systematic investigation of the iteration of these quadratic functions, it will be helpful to have some terminology to describe what happens to points x under iteration of a function f.

A point u is a *fixed point* if $f(u) = u$. We can indicate this by $u \to u$. A fixed point is an *attractor* if all nearby points move towards it under iteration; it is a *repeller* if all nearby points move away from it. For example, 0 and

1 are fixed points for $f(x) = x^2$. The point 0 is an attractor, since starting with initial value $-1 < x_0 < 1$, the sequence of iterates converges to 0. The point 1 is a repeller, since starting with an initial value $0 < x_0 < 1$, the sequence of iterates coverges to 0, and starting with an initial value $1 < x_0$ the sequence of iterates diverges to $+\infty$.

A collection of points $u_1, u_2, \dots, u_k$ forms a *k-cycle* if

$$f(u_1) = u_2,\ f(u_2) = u_3,\ \dots,\ f(u_k) = u_1$$

We can indicate this by

$$u_1 \to u_2 \to u_3 \to \dots \to u_k \to u_1$$

This is also called a cycle of *period k*. A cycle of period one is a fixed point. As for fixed points, a k-cycle can be attracting or repelling depending on whether nearby points move towards it or away from it under iteration. Points near an attracting cycle will start swirling around the cycle, getting closer and closer to it as time goes on. A point u is *preperiodic* if it eventually ends up in some cycle. For example, 1 is preperiodic for the function $f(x) = x^2 - 1$.

14.3 Iterating a quadratic function

We will start with quadratic functions of the form

$$f(x) = ax(1 - x),$$

where the parameter a is between 0 and 4. It turns out that these quadratics are as good (or bad) as any.

We will use two programs to examine this family of quadratic functions. The first is a simple modification of the program in Chapter 1 for iterating linear functions.

Program outline: Iterquad

```
Input: the value of a (0 <=a <=4), the initial value x0,
       and the number N of iterations
Output: the N iterates

INITIALIZE x as an n-dimensional vector
x(1) := x0          %set first entry of x to x0
FOR i = 1 to N
     x(i+1) := a*x(i)*(1-x(i))  %iterate
END FOR
OUTPUT x
```

The second program gives a visual picture of iteration. First try this process by hand: On a piece of graph paper, draw a careful graph of $y = f(x)$ for $a = 1$. Add to this the graph of $y = x$. Choose a value for x_0 on the x-axis. Draw a vertical line from this point to the graph of $f(x)$, draw a horizontal line from this point to the line $y = x$, draw a vertical line from this point to the graph of $f(x)$, then draw a horizontal line to the line $y = x$, and so forth. Convince yourself that this geometric process is indeed just iterating the function. Does your picture resemble a staircase or a cobweb? Compare your graph with the graph that other students get for different choices of x_0. What if you use $a = 0.5$? How about $a = 4$?

The program ITERGRAPH draws a picture like the ones you have just drawn.

Program outline: Itergraph

```
Input: the value of a (0 <=a <=4), the initial value x0,
       and the number N of iterations
Output: a graphical representation of the iteration

RUN Iterquad to get vecter of iterates x

%Create vectors to plot function
xp := [0,0.1,0.2,...,0.9,1] % x values
yp := a * x .* (1-x)        % y values
```

```
INITIALIZE xc as an (2N-1)-dimensional vector
INITIALIZE yc as an (2N-1)-dimensional vector
%xc and yc will be used to plot the cobweb
xc(1) := x(1);
yc(1) := 0;
FOR i = 1 TO N-1
   xc(2*i) := x(i)
   yc(2*i) := x(i+1)
   xc(2*i+1) := x(i+1)
   yc(2*i+1) := x(i+1)
END FOR

PLOT (xp,yp)    %plot function
PLOT (xc,yc)    %plot cobweb on same axes
```

Notice that we use the operation `.*` when we compute `yp` in the code above. The symbol .* indicates element-wise vector multiplication. For example `xp .* yp` would produce a vector whose first entry is the product of the first entries of `xp` and `yp`, second entry is the product of the second entries of `yp` and `yp`, and so on.

Use the programs ITERQUAD and/or ITERGRAPH to help answer the following questions.

Question 1: Choose $a = 1.6$ and try different choices for the initial value x_0. Can you find an attracting fixed point? A repelling fixed point? Which initial values converge to the attracting fixed point? Which go off to infinity?

Question 2: Repeat the previous question with $a = 2$.

Question 3: Find a value of a which gives a 2-cycle. Does this behavior depend on the initial value? Can you find many values for a which result in this behavior?

Question 4: Can you find values of a which give k-cycles for different values of k? Which values of k seem to appear? In each case, does the choice of initial value matter?

(Try values of a between 3.4 and 3.6, as well as values between 3.6 and 4.)

Question 5: If a value of a gives a cycle, is it always an attracting cycle? Can you find any repelling cycles? Why are they hard to find?

Question 6: Find a function which gives a sequence which fails to exhibit any regular behavior (i.e. is "chaotic") even after a very large number of iterations. What, if any, is the effect of the initial value?

Question 7: Can you give an analytic argument to support any of your findings? Here are some suggestions to get you started. Sketch the graphs of these quadratics for $a = 1$, $a = 2$ and a few other choices of $a > 0$. How does the graph change as you vary a? What happens to the zeroes of these functions? The critical points? Next, find the fixed points of $f(x) = ax(1-x)$ in terms of a. Slope was important for the analysis of the iteration of linear functions, so consider the derivative of $f(x)$. For which values of a would you expect convergence to each of your fixed points? Does this behavior depend on the initial value? Which initial values result in convergence to which fixed point? Which result in divergence? Can you prove any of your findings?

14.4 The Feigenbaum diagram

In the previous section you analyzed the behavior of the function $ax(1-x)$ for various individual values of a. We will now do this in a more organized fashion by looking at the long-term behavior of the iterates all at once. The following program starts with a value of a and iterates the function $ax(1-x)$, ignoring the first fifty iterates $x_1, x_2, \dots, x_{50}$ and printing the next fifty iterates $x_{51}, x_{52}, \dots, x_{100}$ on the vertical axis above that value of a. The program then advances to the next value of a. (The parameter a varies between 1 and 4.) The picture that this program produces is called the "Feigenbaum diagram".

At the end of this chapter we have included a two other *MATLAB/Octave* programs, but we omit their pseudocode because of complexity. The program ZOOMDIAGRAM produces the Feigenbaum diagram and allows the user to use a mouse to zoom in on portions of the diagram. The program QUADGRAPHS plots the Feigenbaum diagram, a plot of iterates, and the cobweb plot. When the user clicks on the Feigenbaum diagram, it re-plots the iterates and cobweb using the paramater value a and initial condition from the location of mouse click on the Feigenbaum diagram.

Program outline: Diagram

```
Input: none
Output: the Feigenbaum diagram for f(x) = ax(1-x),
        0 <= a <= 4

a := [0,0.001,0.002,...,3.999,4]  %vector of a values
x := rand(4000,1)    %a vector of random initial x values
                     %one of for each a value

%Compute 1000 iterates to eliminate transients
FOR i = 1 TO 1000
    x := a .* x .* (1-x)  %iterate vector of x with vector
                          %of a values
END FOR

INITIALIZE data as a 100 x 4000 array of zeros
data(1,:) := x %replace first row of data array b x

FOR i = 1:99
    %iterate using ith row of data array to compute i+1 row
    data(i+1) := a.*data(i).*(1-data(i))
END FOR

PLOT (a,data)  %plot values in data array vs. a values
```

Question 8: On the left side of the Feigenbaum diagram there is a single curve. This represents the fact that the function has a single attracting fixed point. When the curve splits into two curves, the function has a two-cycle, and the point jumps back and forth. What is the motion of the point when the curve splits into four curves? Into eight? Can you describe the limiting set of this bifurcation, and the motion of the point on it?

Question 9: On the right side of the Feigenbaum diagram you should be able to see a "period three window", a place where there is a white band crossed by three curves. Can you find any other windows? What are their periods? Is there any pattern here? What is the relation between these windows and the k-cycles from Question 4?

14.5 Examining chaos

Iteration of the function $f(x) = 4x(1-x)$ produces chaotic behavior. Iterate this function using almost any starting value, and you'll see no pattern in the output. It certainly looks like chaos! Although there is no mathematical definition of chaos, there are some properties of chaotic systems that are more precise than just saying that there's no pattern. These properties are sensitive dependence on initial conditions, and some underlying order. We'll explore them in this section.

14.5.1 Sensitive dependence on initial conditions

A chaotic system has the following property: No matter how close two initial values are, they eventually diverge. We can explore this by modifying the program ITERQUAD to do two initial values at once (see ITERQUAD2 below).

Program outline: Iterquad2

```
Input:  two initial values, x01 and x02, between 0 and 1,
        the paramater a, the number N of iterations
Output: two sets of iterates, the absolute difference,
        a plot of the iterates and their difference

RUN Iterquad with x0=x01 to get x1
RUN Iterquad with x0=x01 to get x2
differ := abs(x1-x2)

PLOT x1
PLOT x2
PLOT differ
OUTPUT x1 x2 differ
```

Question 10: Use Iterquad2 to start with two initial values that agree to one decimal place; do they eventually diverge? Now take two initial values that agree to two decimal places, and so forth. What happens?

Question 11: Give a precise mathematical description of "sensitive dependence on initial conditions."

14.5.2 Chaos is not random: histograms

Now we come to a surprising feature of iteration of $4x(1-x)$: Although the sequence of iterates of a typical point looks completely unpredictable, it turns out that this sequence is not at all random!

Question 12: What does it mean to say that a sequence of numbers $x_1, x_2, \ldots$ is random?

We will investigate how well the iterates of $4x(1-x)$ move around the unit interval. For instance, starting with an initial value between 0 and 1 and iterating it, how often is the result less than 1/2? Greater than 1/2? The program RANDOM helps answer this question.

Question 13: Try the program RANDOM starting with different initial values. (Take N to be large, say 1000.) Is there any difference between the number of iterates between 0 and 1/2, and the number of iterates between 1/2 and 1? Are these results consistent with randomness?

Program outline: Random

```
Input: initial value of x between 0 and 1, number N of
       iterations of f(x) = 4x(1-x)
Output: the number of iterates < 1/2, the number >= 1/2

RUN Iterquad to get iterates x
INITIALIZE freq as a 2-dimensional vector of zeros
freq(1) := count of x_i <= 1/2
freq(2) := count of x_i > 1/2

OUTPUT freq
```

Question 14: Now modify the program to count the number of iterates between 0 and 1/3, 1/3 and 2/3, and 2/3 and 1. (You will have to add some more lines to the program.) Repeat the same experiment. Are there any differences?

Try other divisions of the interval $[0, 1]$, for instance into ninths. Do more patterns occur?

Question 15: Among the various divisions of $[0, 1]$ into intervals, is it true that some interval doesn't contain any iterates, even after a large number of iterations? If this never happens (i.e., if all intervals are eventually hit by some iterate), then the system is called "transitive".

Question 16: From the first part of this chapter you may remember some other values of a for which the function $f(x) = ax(1-x)$ seemed to produce chaotic behavior. Repeat Questions 13–15 for these values of a.

14.5.3 Chaos is not random: repelling periodic points

There is another way in which the function $f(x) = 4x(1-x)$ is not random: It has lots of periodic points! Of course 0 and 1 are fixed points, but these are rather boring. A surprise is that the point $[\sin(\pi/5)]^2$ has period 2.

Exercise 1: Verify this. (You have to use the exact value $[\sin(\pi/5)]^2$ and not a decimal approximation; a computer algebra program makes this task easier.)

Why didn't we see this before? The answer is that it is a repelling periodic point, so any finite decimal approximation which we could feed into a computer eventually gets thrown off.

Exercise 2: Using a calculator, find the decimal expansion of $[\sin(\pi/5)]^2$ to as many places as you can, and calculate its iterates using f. For how many iterations does it look like a cycle of period two?

There are many other repelling periodic points, of all possible periods, and in fact they form a "dense" subset of the interval. In the next section we'll get some ideas about how to find them.

14.6 The tent and sawtooth functions

Two other functions which have chaotic behavior like that of $f(x) = 4x(1-x)$ are the "tent function" and the "sawtooth function."

The tent function $T(x)$ is defined by

$$T(x) = \begin{cases} 2x & \text{if } 0 \leq x \leq 1/2 \\ 2(1-x) & \text{if } 1/2 \leq x \leq 1 \end{cases}$$

The sawtooth function $S(x)$ is defined by

$$S(x) = \begin{cases} 2x & \text{if } 0 \leq x < 1/2 \\ 2x-1 & \text{if } 1/2 \leq x \leq 1 \end{cases}$$

Question 17: Sketch a graph of the tent function. Why is it called the tent function? (This function is also called "stretching and folding," which is what happens when dough is kneaded.)

Question 18: Show that the tent function exhibits sensitivity to initial conditions (Think of it as kneading.)

Question 19: It is easy to work with the tent function by hand: Find periodic points of period 2, 3, 4, 5 and 6 by choosing some simple fractions as initial points.

Question 20: Sketch a graph of the sawtooth function. The sawtooth function is called "stretch, cut and paste"; can you see why?

Question 21: Show that the sawtooth function exhibits sensitivity to initial conditions.

Question 22: Find all periodic points for the sawtooth function. (Hint: One way of understanding the sawtooth function is to write numbers in their binary representation.)

Question 23: The relation between the tent and sawtooth functions is given by

$$T(T(x)) = T(S(x))$$

for any $0 \leq x \leq 1$. Verify this.

Question 24: Use this relation to find all periodic points for the tent function.

14.7 Conjugacy

It's relatively easy to find periodic points for the tent and sawtooth

functions. On the other hand, it is hard to find periodic points for the function $f(x) = 4x(1-x)$ because they can't be expressed with finite decimals. If there were a way to translate back and forth between the difficult function and the easier ones, it would lead us to these elusive periodic points.

This translation back and forth is provided by the notion of conjugacy. In fact, if we let

$$h(x) = \sin^2(\pi x/2)$$

then

$$f^k(h(x)) = h(T^k(x))$$

for any k. Thus iterating f is equivalent to iterating T, which is much easier to understand. You can read more about this in *Fractals for the Classroom* by Peitgen et al., Chapter 10.

14.8 Iterating other functions

There are some other functions we could look at rather than the function $f(x) = ax(1-x)$. If you are working with enough people, divide into groups, each group taking one of the functions below. Compare and contrast their behavior under iteration to the function $ax(1-x)$.

1. $a - (x - \sqrt{a})^2$
2. $x^2 - a$
3. $a \sin(x)$
4. $x^4 - a$

14.9 Listening to chaos

You can easily modify the first program of this chapter so that rather than printing the value of an iterate x, the program generates a note with pitch x. (For details on how to do this, see the specific programs at the end of this chapter.) What does does an attractor sound like? A two-cycle? A four-cycle? Chaos? (Drive your classmates crazy!)

Program outline: Music

```
Input: Number of iterates N, parameter a
Output: the sound of iterates of f(x) = ax(1-x)
RUN Iterquad to get iterates x
FOR i = 1 TO N
     Play the note with pitch x for a short period of time
     Brief moment of silence
     PLOT x
     PLOT (i,x) %plot a point showing current note
END FOR
```

14.10 Computer programs

14.10.1 *MATLAB/Octave* programs

Program: Iterquad

```
function x=iterquad(a,x0,N)
%function x=iterquad(a,x0,N)
%
%Iterates f(x)=ax(1-x)
%
%Inputs
%  a - parameter in function (0<a<=4)
%  x0 - initial condition (0<=x0<=1),
%  N - number of iterations
%Output
%  x - vector of iterates

x=zeros(1,N); %ititialize x as a vector of zeros
x(1)=x0;      %set initial condition

for i=1:(N-1)
  x(i+1)=a*x(i)*(1-x(i)); %iterate
end
```

We can plot the result and label the axes using:

```
>> plot(x, 'k-o') %plot
>> xlabel('i')   %label axes
>> ylabel('x_i')
```

Program: Itergraph

```
function x=itergraph(a,x0,N)
%function x=itergraph(a,x0,N)
%
%Iterates f(x)=ax(1-x)
%
%Inputs
%  a - parameter in function (0<a<=4)
```

```
%  x0 - initial condition (0<=x0<=1),
%  N - number of iterations
%Output
%  x - vector of iterates
%  generates a cobweb plot

x=iterquad(a,x0,N); %get list of iterates

xp=linspace(0,1); %vector of x for plotting
fp=a*xp.*(1-xp);  %vector of y for plotting

xc=zeros(2*N-1,1);  %y coordinates for cobweb
yc=zeros(2*N-1,1);  %x coordrnites for cobweb
xc(1)=x(1);
yc(1)=0;
for i=1:N-1
  xc(2*i)=x(i);
  yc(2*i)=x(i+1);
  xc(2*i+1)=x(i+1);
  yc(2*i+1)=x(i+1);
end; %for


plot(xp,fp,'b-',[0,1],[0,1],'k-',xc,yc,'g-') %create plot
%label axes
xlabel('x_i')
ylabel('x_{i+1}')
set(gca,'FontSize',16);
axis('square');
```

Program: Diagram

```
function []=diagram(amin,amax,xmin,xmax,numtrans,numpoints)
%function []=diagram(amin,amax,xmin,xmax,numtrans,numpoints)
```

```
%Creates Feigenbaum Diagram for f(x)=ax(1-x)
%
%Inputs (optional can be called with no inputs)
%  amin - min 'a' value (default = 0)
%  amax - max 'a' value (default = 4)
%  amin - min x value to plot (default = 0)
%  amin - max x value to plot (default = 1)
%  numtrans - # of steps taken to eliminate transients (default=1000)
%  numpoints - # of points for attractor (default=100)
%Output
%  none - creates plot

%if no input arguments given set defaults
if nargin == 0
   amin=0;
   amax=4;
   xmin=0;
   xmax=1;
end; %if

if nargin<6
   numpoints=100;
   if nargin<5
      numtrans=1000;
   end; %if
end; %if

numpoints=100; %number of points to represent long term dynamics
N=4000;   %number of entries in 'a'
a=linspace(amin,amax,N);    %vector of 'a' values in vector
data=zeros(numpoints,N); %initialize matrix of points for diagram
x=rand(1,N);

for i=1:numtrans %run numtrans steps to eliminate transients
   x=a.*x.*(1-x);
end; %for

data(1,:)=x;
```

```
%iterate an additional "numsteps" and save in data matrix
for i=1:(numpoints-1)
   data(i+1,:)=a.*data(i,:).*(1-data(i,:));
end; %for

%plot result
plot(a,data','b.','MarkerSize', 1)
axis([amin,amax,xmin,xmax])
xlabel('a')
ylabel('x')
set(gca,'FontSize',16);
```

Notice that this program has a set of optional inputs not described in the pseudocode. These inputs are clearly described in the code.

Program: Zoomdiagram

```
%Script file
%Creates Feigenbaum Diagram for a*x(1-x) which allows for zooming
%and recalculation. Be aware that code runs slow, wait for plus
%to appear after clicking.

pause on;
go=1;
amin=2.5;
amax=4;
xmin=0;
xmax=1;
while go
   %increase number of iterations as window becomes smaller
   numtrans=round(29000*(1-(xmax-xmin)^.2)+1000);
   numpoints=round(min(95+5*(xmax-xmin)^(-4),1e+12));
   diagram(amin,amax,xmin,xmax,numtrans) %plot diagram
   title('Click corner of box to zoom or right click to quit')
   [c1a,c1x,button]=ginput(1); %get single mouse click
```

```
   if button~=1; %right click?
      go=0;
   else
      hold on
      plot(c1a,c1x,'r+','MarkerSize',15) %place plus at corner
      title('Click other corner of box to zoom')
      [c2a,c2x,button]=ginput(1); %another mouse click
      plot(c2a,c2x,'r+','MarkerSize',15) %place plus at other corner
      pause(1)
      amin=min(c1a,c2a);
      amax=max(c1a,c2a);
      xmin=min(c1x,c2x);
      xmax=max(c1x,c2x);
      hold off
   end; %if
end; %while
title('Program is done')
```

Program: Quadgraphs

```
%Script file
%Creates Feigenbaum Diagram for a*x(1-x) along with cobeb plot and
%the iterates. Allows user to click on Feigenbaum diagram to see
%cobweb plot and iterates with initial condition and parameter
%value corresponding to the location of mouseclick  on Feigenbaum
%diagram.

subplot(2,2,1:2) %top graph
diagram %create Feigenbaum diagram
go=1; %variable to conntinue while loop
a=2;  %inital choice of parameter
x0=0.1; %inital coice for initial condition
N=20;  %number of iterates to plot
while go
```

```
   title('Click Feigenbaum Diagram or right click to quit')
   subplot(2,2,3) %bottom left graph
   x=itergraph(a,x0,N); %create cobweb plot
   title(['a=',num2str(a)])
   subplot(2,2,4) %bottom right graph
   plot(x,'b-o') %plot iterates
   axis([1,N,0,1])
   xlabel('i')
   ylabel('x_i')
   set(gca,'FontSize',16);
   title(['a=',num2str(a)])
   [a,x0,button]=ginput(1); %get mouse click
   if button~=1;
      go=0;
      title('Program is done')
   end; %if
end; %while
```

Program: Iterquad2

```
function [x1,x2,differ]=iterquad2(x01,x02,a,N)
%function [x1,x2,differ]=iterquad2(x01,x02,a,N)
%
%Iterates f(x)=ax(1-x) for 2 different initial conditions.
%Plots trajectories and absolute difference
%
%Inputs
%  x01 - initial condition 1 (0<=x01<=1),
%  x02 - initial condition 2 (0<=x01<=1),
%  a - parameter in function (0<a<=4, default a=4)
%  N - number of iterations (default N=20)
%Output
%  x1 - vector of iterates solution 1
%  x2 - vector of iterates solution 2
```

```
% differ - vector of absolute difference of solutions |x1-x2|

if nargin<4 %function called with fewer than 4 inputs
   N=20; %default N
   if nargin <3 %function called with fewer than 3 inputs
      a=4; %default 'a'
   end; %if
end; %if


x1=iterquad(a,x01,N,0); %solve with intial condition x01
x2=iterquad(a,x02,N,0); %solve with intial condition x02
differ=abs(x1-x2); %absolute difference
ivals=0:N-1; %i values

%Plot x_1
subplot(3,1,1)
plot(ivals,x1,'k-o')
axis([0,N-1,0,1])
ylabel('x_1')


%Plot x_2
subplot(3,1,2)
plot(ivals,x2,'k-o')
axis([0,N-1,0,1])
ylabel('x_2')


%Plot absolute difference
subplot(3,1,3)
plot(ivals,differ,'k-o')
axis([0,N-1,0,1])
ylabel('|x_1-x_2|')
xlabel('i')
```

Program: Random

```
function [freq]=random(x0,N,a)
%function []=random(x0,N,a)
%
%Creates histogram of iterates of ax(1-x)
%
%Inputs
%  x0 - initial condition (0<=x0<=1),
%  N - number of iterations
%  a - parameter in function (0<a<=4, default=4)
%Output
%  freq - vector with count number of iterates
%         less or equal to 0.5 and greater than 0.5

if nargin<3 %if "a" not given, set a=4
   a=4;
end; %if

x=iterquad(a,x0,N,0); %run iterquad
freq=zeros(1,2);      %vector of frequencies
freq(1)=sum(x<=0.5);  %count less or equal 1/2
freq(2)=sum(x>0.5);   %count greater than 1/2
```

Program: Music

```
function []=music(N,a)
%function []=music(N,a)
%
%Plays notes correpsonfing to iterates of ax(1-x)
%with random initial condition, if called with
%a=-1 a random sequence is generated.
```

```
%
%Inputs
%  N - Number of notes (default = 20),
%  a - parameter, if a=-1 plays random notes
%      otherwise 0<a<=4, default=4
%Output
%  none - plays sound with plot

pause on
if nargin<2
   a=4; %parameter
   if nargin<1
     N=20;
   end; %if
end; %if
s = 0.5; %0.5 second tone
sr = 44100; %samplng rate
t = linspace(0, s, sr*s); % time interval
%function to attenuate note slightly at end
attenuate=1.5*(t/(sr*s)).^0.1.*(1-t/(sr*s)).^0.8;
low=200; %frequency of lowest note played
high=800; %frequency of lowest note played

x0=rand;
y=zeros(length(t),N);
if a==-1
   x=rand(1,N);
else
   x=iterquad(a,x0,N);
end; %if

for i=1:N
   %note correspnding to x
   y = attenuate.*(sin((low+(high-low)*x(i))*2*pi*t));
   plot(1:N,x,'b-o',i,x(i),'r*') %plot iterates red at note
   axis([1,N,0,1])
   xlabel('i')
   ylabel('x_i')
```

```
    set(gca,'FontSize',16);
    pause(0.1) %pause to allow plot to come up
    sound(y, sr); %play sound
end; %for
```

14.10.2 *Mathematica* programs

The Basic Program for Quadratic Iteration

The following program performs quadratic iteration, using the function $f(x) = ax(1-x)$, much as the linear iteration program in Chapter 1 worked. You must run this program before you do anything else in this notebook; everything else runs independently of one another.

IterQuad[a_, n_, x0_]:=NestList[Function[x, $ax(1-x)$], x0, n]

Here is an example where $a = 1.5$, the number of iterations $= 10$, and the initial value $x_0 = 0.1$. **Always set a to a number between 0 and 4 and x_0 to a number between 0 and 1.**

IterQuad[1.5, 10, .1]

$\{0.1, 0.135, 0.175163, 0.216721, 0.254629, 0.28469, 0.305462, 0.318233,$

$0.325441, 0.329294, 0.331289\}$

Cobweb Diagrams

The following program will display a "cobweb" diagram for quadratic iteration, with the same inputs as IterQuad:

```
ItPoints[itlist_]:=Partition[Drop[Drop[Flatten[Transpose[
    {itlist, itlist, itlist, itlist}]], 1], -1], 2];
IterGraph[a_, n_, x0_, pr_ : {0, 1}]:=Show[ListPlot[ItPoints[IterQuad[a, n, x0]],
    Joined -> True, PlotRange -> {pr, pr}], Plot[{x, ax(1 - x)}, {x, 0, 1},
    PlotStyle -> {{Red, Thick}, {Green, Thick}}]]
IterGraphEnd[a_, n_, x0_, k_ : 16]:=Show[ListPlot[Drop[ItPoints[
    IterQuad[a, n, x0]], 2(n - k)], Joined -> True, PlotRange -> {{0, 1}, {0, 1}}],
    Plot[{x, ax(1 - x)}, {x, 0, 1}, PlotStyle -> {{Red, Thick}, {Green, Thick}}]]
```

Here is an example, where $a = 2.6$, $x_0 = 0.9$ and the number of iterations is 20.

```
IterGraph[2.6, 20, .9]
```

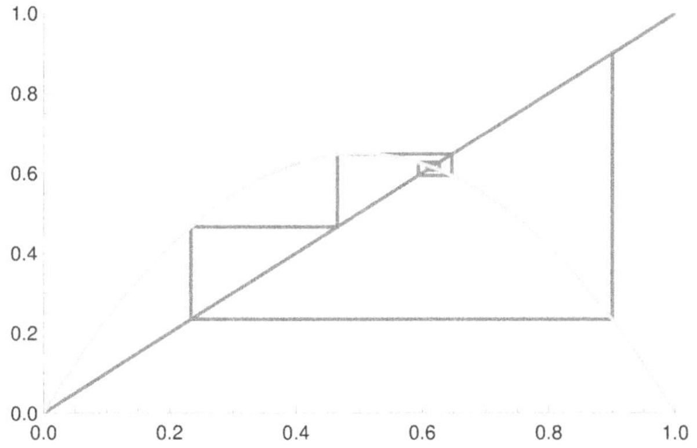

You can change the viewing window by an optional fourth argument. The following command graphs the portion of the graph in the square $[0.5,0.7]\times[0.5,0.7]$.

IterGraph[2.6, 20, .9, {.5, .7}]

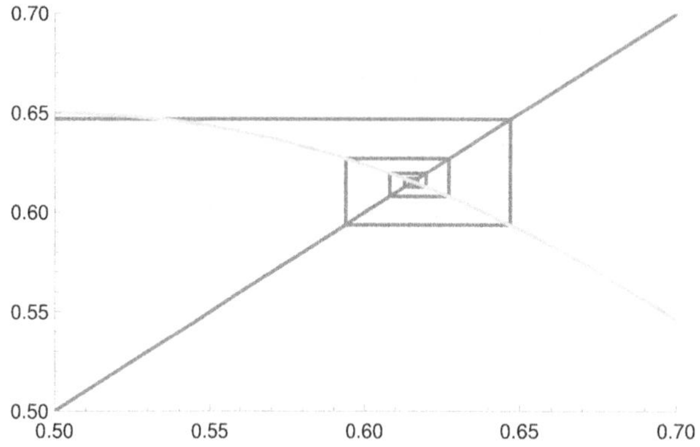

The next command is similar to IterGraph but only shows the last sixteen points on the diagonal. Here is an example displaying an attractive 4-cycle ($a = 3.51$):

IterGraphEnd[3.51, 100, .9]

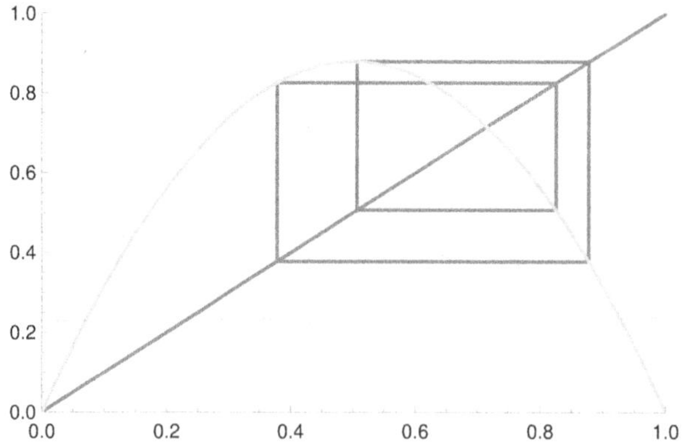

Confirming that this is an example displaying an attractive 4 - cycle:

IterQuad[3.51, 100, .5]

{0.5, 0.8775, 0.377303, 0.824659, 0.507535, 0.877301, 0.377831, 0.825112, 0.5065, 0.877352, 0.377696, 0.824996, 0.506764, 0.877339, 0.377729, 0.825024, 0.5067, 0.877342, 0.377721, 0.825018, 0.506716, 0.877342, 0.377723, 0.825019, 0.506712, 0.877342, 0.377722, 0.825019, 0.506713, 0.877342, 0.377722, 0.825019, 0.506713, 0.877342, 0.377722, 0.825019, 0.506713, 0.877342, 0.377722, 0.825019, 0.506713, 0.877342, 0.377722, 0.825019, 0.506713, 0.877342, 0.377722, 0.825019, 0.506713, 0.877342, 0.377722, 0.825019, 0.506713, 0.877342, 0.377722, 0.825019, 0.506713, 0.877342, 0.377722, 0.825019, 0.506713, 0.877342, 0.377722, 0.825019, 0.506713, 0.877342, 0.377722, 0.825019, 0.506713, 0.877342, 0.377722, 0.825019, 0.506713, 0.877342, 0.377722, 0.825019, 0.506713, 0.877342, 0.377722, 0.825019, 0.506713, 0.877342, 0.377722, 0.825019, 0.506713, 0.877342, 0.377722, 0.825019, 0.506713, 0.877342, 0.377722, 0.825019, 0.506713, 0.877342, 0.377722, 0.825019, 0.506713, 0.877342, 0.377722, 0.825019, 0.506713}

The Take command with -5 in the end gives you the last 5 iteration values:

Take[IterQuad[3.51, 100, .5], −5]

{0.506713, 0.877342, 0.377722, 0.825019, 0.506713}

Feigenbaum Diagram

This program, which takes no input, will produce the Feigenbaum diagram. The Feigenbaum diagram only shows attractive fixed points and cycles.

```
FeigDiagram:=Module[{newones, allofthem = {}},
Do[newones = Transpose[{Table[a, {100}], Take[IterQuad[a, 200, .1], −100]}];
    allofthem = Join[allofthem, newones],
    {a, 0, 4, .015}];
    ListPlot[allofthem, AxesLabel → {a, None}, ImageSize → 500]];
FeigDiagram
```

Sensitive Dependence on Initial Conditions

The following program performs quadratic iteration, with $a = 4$ fixed, on two initial inputs, x_1 and x_2. (To change a, simply replace both occurrences of "4" to whatever value you like.)

```
IterQuad2[n_, x1_, x2_]:=MatrixForm[Transpose[{Range[n + 1],
IterQuad[4, n, x1], IterQuad[4, n, x2]}]]
```

Here is an example where the number of iterations is 25, and the initial conditions are $x_1 = 0.1$ and $x_2 = 0.1001$.

IterQuad2[25, .1, .1001]

1	0.1	0.1001
2	0.36	0.36032
3	0.9216	0.921958
4	0.289014	0.287806
5	0.821939	0.819895
6	0.585421	0.590669
7	0.970813	0.967116
8	0.113339	0.12721
9	0.401974	0.444109
10	0.961563	0.987505
11	0.147837	0.0493563
12	0.503924	0.187681
13	0.999938	0.609828
14	0.000246305	0.951752
15	0.000984976	0.183682
16	0.00393603	0.599772
17	0.0156821	0.960183
18	0.0617448	0.152928
19	0.23173	0.518164
20	0.712124	0.99868
21	0.820014	0.00527196
22	0.590364	0.0209767
23	0.967337	0.0821467
24	0.126384	0.301594
25	0.441645	0.842541
26	0.986379	0.530663

Chaos is not Random: Histograms

The following program will perform quadratic iteration **(with a=4 fixed)**, and then report how many of the iterates are in different parts of the interval from 0 to 1. (To change a, simply replace the "4" to whatever value you like.)

```
RandomIter[n_, x0_, m_]:=Module[{i, iq = Drop[IterQuad[4, n, x0], 1], b, b2},
b = BinCounts[iq, {0, 1 + 1/m, 1/m}];
b2 = Join[Drop[b, -2], {b[[m]] + b[[m + 1]]}];
Do[Print["The number of values between ", (i - 1)/m, " and ", i/m, " is ", b2[[i]]],
    {i, m}]];
```

Here is an example with 1000 iterations, initial condition $x_0 = 0.768$ and the interval from 0 to 1 being divided into sevenths.

RandomIter[1000, .768, 7]

The number of values between 0 and $\frac{1}{7}$ is 231

The number of values between $\frac{1}{7}$ and $\frac{2}{7}$ is 117

The number of values between $\frac{2}{7}$ and $\frac{3}{7}$ is 109

The number of values between $\frac{3}{7}$ and $\frac{4}{7}$ is 85

The number of values between $\frac{4}{7}$ and $\frac{5}{7}$ is 97

The number of values between $\frac{5}{7}$ and $\frac{6}{7}$ is 101

The number of values between $\frac{6}{7}$ and 1 is 260

RandomIter[10, 1/2, 2]

The number of values between 0 and $\frac{1}{2}$ is 9

The number of values between $\frac{1}{2}$ and 1 is 1

The following command will give the same information displayed as a histogram:

```
RandomIterH[n_, x0_, m_]:=Module[{i, iq = Drop[
IterQuad[4, n, x0], 1], b, b2}, b = BinCounts[iq, {0, 1 + 1/m, 1/m}];
b2 = Join[Drop[b, -2], {b[[m]] + b[[m + 1]]}];
z = Table [Row [{"[", (i-1)/m, ", ", i/m, ")"}] , {i, m}] ;
z2 = Join [Drop[z, -1], {Row [{"[", 1 - 1/m, ",", 1, "]"}]}] ;
BarChart[b2, ChartLabels -> z2, ImageSize -> 500]]

RandomIterH[1000, .769, 7]
```

Chapter 15

Iterated Linear Maps in the Plane

15.1 Introduction

In this chapter we consider some properties of linear maps in the plane. We will represent linear maps by *matrices*. It turns out that matrices can be multiplied, and iteration of a linear map is just repeated matrix multiplication.

As you work through this chapter you will:

- Learn matrix notation for a linear map;
- Examine typical examples of what happens when you iterate a linear map in the plane; and
- See the emergence of a special direction and number associated with a matrix: the largest *eigenvalue*, and its corresponding *eigenvector*.

The notion of eigenvector of a linear map is an extremely important one. You will encounter it throughout pure and applied mathematics.

15.2 Multiplying matrices

"Map" or "mapping" is a synonym for "function." But when we speak of linear maps in the plane, we mean a more restricted notion than the definition of linear function we used in Chapter 1. A function

$$f(x) = ax + b$$

in this setting is called an *affine* map. A *linear* map is one where the constant term $b = 0$. You already know what happens when you iterate a linear map on the line (think of it as the x coordinate axis). If we write $x_{n+1} = f(x_n)$, as usual, then for a linear map

$$x_{n+1} = ax_n \tag{15.1}$$

The result of iterating a linear map on the line depends entirely on whether the constant a is such that $|a| > 1$, $|a| = 1$, or $|a| < 1$. The sequence of x's that you get either grows in magnitude, stays the same magnitude, or decays in magnitude (assuming you don't start with an initial x which is zero). You can easily picture how this sequence looks on the line.

The situation gets a lot more complicated when you have two input variables and two output variables. In this case a linear function involves *four* constant coefficients instead of one:

$$f(x, y) = (a_{11}x + a_{12}\,y,\ a_{21}x + a_{22}\,y).$$

If we write $(x_{n+1}, y_{n+1}) = f(x_n, y_n)$, then we have

$$\begin{aligned} x_{n+1} &= a_{11}x_n + a_{12}\,y_n \\ y_{n+1} &= a_{21}x_n + a_{22}\,y_n \end{aligned} \tag{15.2}$$

The sequence of (x_n, y_n)'s that you get when you iterate this linear function could be pictured as a sequence of points in the Cartesian coordinate plane, but what does this sequence of points look like? That is the question we are investigating.

Exercise 1: Choose some values for the four constants called a_{ij} in Equation (15.2) and work out a few examples. Plot the sequence of points in the plane (x_0, y_0), (x_1, y_1), (x_2, y_2), ... for each choice you make, starting with some convenient choice of (x_0, y_0).

You will find, of course, that any point (x_n, y_n) in the plane gets mapped to another point (x_{n+1}, y_{n+1}) in the plane, but it may be hard to say anything more illuminating, unless you pick particularly simple values for the constants. Those cases which are simple enough to understand right away are important to know about, though.

Exercise 2: If you haven't already, try the map

$$\begin{aligned} x_{n+1} &= x_n \\ y_{n+1} &= 2\, y_n \end{aligned}$$

(Check your understanding of the notation by identifying each of a_{11}, a_{12}, a_{21} and a_{22} in this case.) Find some more examples of the map given in equation (15.2) which seem simple. Plot typical sequences of points in the plane generated by these simple examples, starting from some convenient (x_0, y_0).

A common notation for the map in equation (15.2) is

$$\begin{pmatrix} x_{n+1} \\ y_{n+1} \end{pmatrix} = \begin{pmatrix} a_{11} & a_{12} \\ a_{21} & a_{22} \end{pmatrix} \begin{pmatrix} x_n \\ y_n \end{pmatrix} \tag{15.3}$$

If you want to know what equation (15.3) means, look at equation (15.2). It means exactly that. If you are unfamiliar with this notation, compare the two versions and practice on some specific examples until you can look at one and translate it into the other easily.

Equation (15.3) is sometimes also written

$$X_{n+1} = AX_n \,. \tag{15.4}$$

In this compressed notation, the X_{n+1} on the left really stands for the *two* quantities x_{n+1} and y_{n+1}. It is a *vector*, which for our purpose here means a column of two numbers:

$$X_{n+1} = \begin{pmatrix} x_{n+1} \\ y_{n+1} \end{pmatrix} .$$

(If we're thinking of X_{n+1} as a *point* in the plane, we'll write (x_{n+1}, y_{n+1}). If we want to save space but emphasize that we're thinking of X_{n+1} as a *column vector*, we'll use square brackets and write $[x_{n+1}, y_{n+1}]$.) Similarly, X_n stands for the column vector consisting of the two numbers x_n and y_n. Finally, A represents the array of four constants in equation (15.3). It is a *matrix*, just the usual name for such an array. We say A has two *rows*: (a_{11}, a_{12}) and (a_{21}, a_{22}), and two *columns*

$$\begin{pmatrix} a_{11} \\ a_{21} \end{pmatrix} \quad \text{and} \quad \begin{pmatrix} a_{12} \\ a_{22} \end{pmatrix}.$$

Notice that the subscripts on the four constants follow a standard pattern: a_{ij} is the number in the ith row and j^{th} column of A. Look back at equations (15.2) and (15.3) to see what equation (15.4) really means.

Exercise 3: Write the map

$$\begin{aligned} x_{n+1} &= x_n + 2y_n \\ y_{n+1} &= 3x_n - y_n \end{aligned}$$

in the form of equation (15.3). Write the map

$$\begin{pmatrix} x_{n+1} \\ y_{n+1} \end{pmatrix} = \begin{pmatrix} 4 & -2 \\ -1 & 1 \end{pmatrix} \begin{pmatrix} x_n \\ y_n \end{pmatrix}$$

in the form of equation (15.2)

Exercise 4. Write your maps from Exercises 1 and 2 in matrix notation.

In equation (15.4) the matrix A appears to "multiply" the vector X_n, and this operation of the constants in A on the two components of X_n really is called *matrix multiplication*. Iterating the linear map is then simply repeated matrix multiplication. This situation looks suggestively like that in equation (15.1), but it can't be that simple. What would be the analogue of the statement $|a| > 1$? The matrix A has four numbers in it, and some might be greater than 1 in magnitude and others less.

15.3 An example to start

The following example is interesting to work through. Let the matrix A be

$$A = \begin{pmatrix} 1/5 & 99/100 \\ 1 & 0 \end{pmatrix}. \tag{15.5}$$

Plot the sequence which starts with $(x_0, y_0) = (1, 1)$. Try it again, but this time starting from $(1, 0)$. Try it again from $(0, 1)$. You will find after enough iterations that for any of these starting points the iteration sequence begins to look the same in a certain sense. The points in the sequence seem to converge onto a line and to move along that line, much like the points in the 1-dimensional case of equation (15.1). Confirm this in the following exercise.

Exercise 5: Using the matrix A from equation (15.5) and a choice of (x_0, y_0), find a sequence of iterates (x_n, y_n) using equation(15.3). Look at the sequence of slopes y_n/x_n to see that the iterates are converging to some definite line of slope m. What is m? Look at the sequence of ratios x_{n+1}/x_n and y_{n+1}/y_n to see that the effect of the map after many iterations is just scaling by a certain constant a:

$$\begin{aligned} x_{n+1} &= a x_n \\ y_{n+1} &= a y_n \end{aligned}$$

What is a? This is just like the 1-dimensional map of equation (15.1), but the line in question is not the x-axis.

If you did Exercise 5 by hand, you are probably ready to computerize the calculation. Below is pseudocode for a program PLANAR1 that does the necessary computations and plots the iterates.

Program outline: Planar1

```
Input: 2 X 2 matrix  A = [a11, a12; a21, a22]
       initial vector X0 = [x0;y0]
       number of iterates n
Output:  n iterates X = A*X and a plot of iterates

INITIALIZE its as an (2 x n+1)-dimensional array
its(:,1) := x0          %set first column of its to X0
FOR i = 1 TO n
   its(i+1) := A*its(:,i) % * inidcates matrix multiplcation
END FOR
PLOT its
OUTPUT its
```

The program prints the ratios suggested in Exercise 5, and also plots the iterated points in the plane, all for the given initial vector $X_0 = [x_0, y_0]$.

Exercise 6: Use the program to confirm your results for A from equation (15.5) and for some of your examples from Exercises 1 and 2, trying it for several choices of X_0.

Another possibility is to choose many initial X_0's randomly in the square $\{(x, y) : |x| \leq 1, |y| \leq 1\}$. and to follow *all* the X_0's for many iterations. By taking many random initial points in the same picture, you can examine how the choice of starting point affects the iteration. PLANAR2 is a program that randomly chooses 17 initial vectors X_0 in the square described above, then finds 25 iterates of each initial vector, and plots all 442 points. (If your programming language permits it, you might like to color each of the 17 sequences of iterates differently.) Note that there is nothing special about the choice of 25 or 17—they're just convenient values.

Program outline: Planar2

```
Input: the 2 X 2 matrix  A = [a11, a12; a21, a22]
Output:  25 iterates of each of 17 different initial
         vectors X0(j) (for j = 1 to m)

m := 17 %number of random inital vectors
n := 25 %number of iterates for each initial condition
FOR j = 1 TO m
   %x0 is a 2x1 vector of random numbers between -1 and 1
   x0 := [2*rand-1; 2*rand-1]
   CALL Planer1 with A, x0, and n
END FOR
```

Exercise 7: Run PLANAR2 and confirm that the result is something like Figure 15.1.

From your results with PLANAR2 you can see graphically that this iteration in the plane eventually looks like iteration on the line. The 17 points, initially chosen randomly, move to a certain line in the plane under iteration of "multiplication" by A, and on that line they move away from the origin as if each iteration were multiplication by a number a slightly greater than 1.

Exercise 8: Here is a way to confirm this by using a procedure that is slightly different from Exercises 5 and 6. If the successive points (x_1, y_1), (x_2, y_2), ... are moving out on a line as if they were being multiplied by $a > 1$, then after a while

$$X_{n+1} = (x_{n+1}, y_{n+1}) = AX_n = (ax_n, ay_n).$$

Suppose after every iteration of the function you were to scale X_{n+1} by a positive factor c. In other words, suppose you replaced X_{n+1} by

$$cX_{n+1} = (cx_{n+1}, cy_{n+1}) = (cax_n, cay_n).$$

If you happen to choose $c = 1/a$, you would get $cX_{n+1} = X_n$, and you would see the points stay *fixed* on iteration. If you were to choose $c > 1/a$, then the

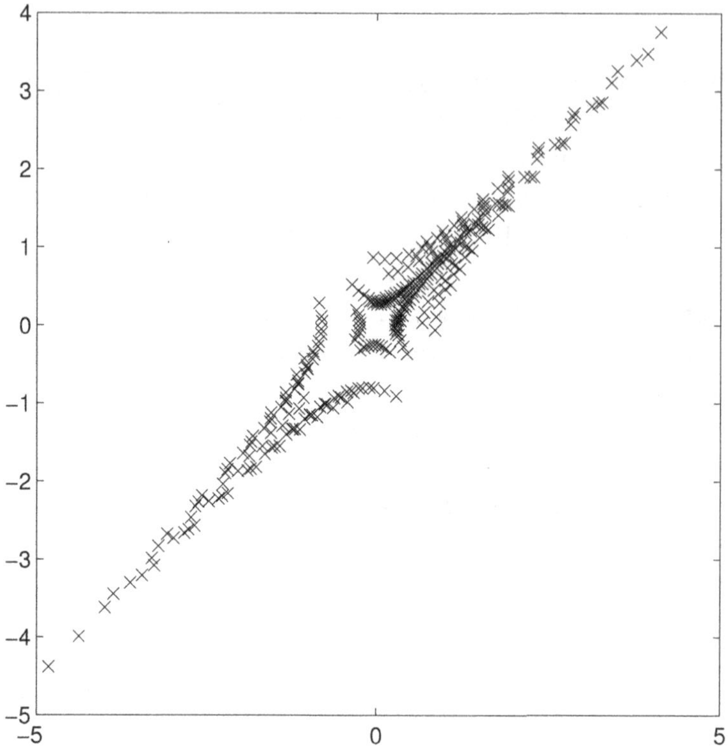

Fig. 15.1: Iterates obtained using matrix A

resulting scale factor ca would be larger than 1, and the new points would move *out*, away from the origin, on iteration. If you chose $c < 1/a$, then the scale factor ca would be less than 1, and the new points would move *in*, toward the origin. So, by trial and error, you might determine the scale factor c that is *just right*—the points neither move away from the origin nor toward it; they stay fixed. Then you could conclude the map itself was multiplying by $a = 1/c$.

You can use the program PLANAR2 to permit you to determine a in this trial and error way. Instead of inputting the matrix A, input $c * A$. Keep trying different values for c until you see the initial points stay fixed (not move out or in). The c that works in this way should be the reciprocal of the number a you found in a different way in Exercises 5 and 6.

The number a is called the *largest eigenvalue* of A, and the line $y = mx$ along which the points eventually move on iteration determines a corresponding *eigenvector* of A of the form $X = [u, mu]$. Notice that the eigenvector is not unique; only the "direction" m is unique. (Where is an eigenvector X in Figure 15.1, for example?)

> *Linguistic aside:* The word "eigen" in German means "own," as in one's own. The mathematical terms *eigenvalue* and *eigenvector* indicate that the value a and a vector determined by the line $y = mx$ are distinctively associated with the linear map described by the matrix A. Sometimes the terminology "characteristic value" and "characteristic vector" is used in English, but eigenvalue and eigenvector are the standard terms.

This means that iteration of the linear map described by A is eventually simple: after many iterations, it is just like the 1-dimensional case, only the effective multiplier a is not obvious, and the line $y = mx$ is also not obvious, just from knowing A.

It is interesting to repeat the same steps with the matrix B given by

$$B = \begin{pmatrix} 0 & 1 \\ 100/99 & -20/99 \end{pmatrix}$$

This matrix is the *inverse* of A, denoted A^{-1}. If $X_1 = AX_0$, then $X_0 = BX_1$. Multiplying by B just undoes the effect of multiplying by A.

Exercise 9: Try this out for $X_0 = [x_0, y_0]$ and confirm that multiplication by B undoes multiplication by A.

What happens when you iterate with B? If iterating A just produces motion along a certain line, won't B just produce motion backward along the same line, since it undoes the action of A? This is in a sense true, but it is not what you see when you try it. Rather you will see that B produces expansion (i.e., multiplication by a number greater than 1) along a *different* line. That is, B has its own largest eigenvalue, and corresponding direction (eigenvector), different from the one found before for A. The situation is as illustrated in Figure 15.2.

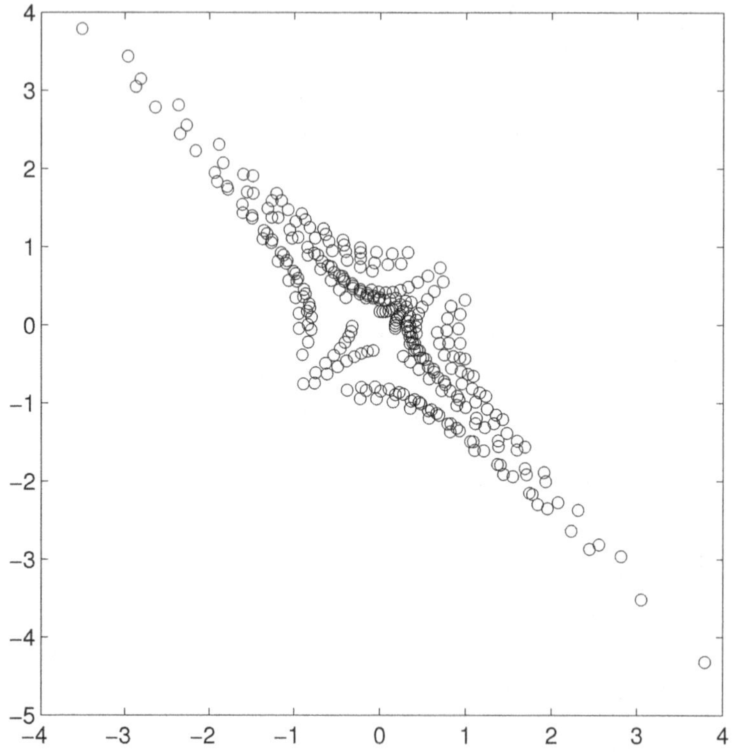

Fig. 15.2: Iterates obtained using matrix B

Exercise 10: Find the largest eigenvalue and a corresponding eigenvector for the matrix B. (Note: "largest" here must be understood "in magnitude." It is actually a negative number—compare with the case of a linear map on the line when $a < 0$.)

15.4 Questions to explore

Use the computer or hand calculations to investigate the following questions.

Question 1: Try some matrices of your own, perhaps making small changes in the matrices A or B to start with. Do you always find the same qualitative behavior that we observed in the examples?

Question 2: Consider the "simple" maps that you invented in Exercise 2. Can you say anything about their largest eigenvalues and corresponding eigenvectors?

Question 3: Here is the matrix of a linear map with a behavior quite different from our starting example:

$$A = \begin{pmatrix} 1 & -1 \\ 1 & 1 \end{pmatrix}$$

The scaling method suggested in Exercise 8 still simplifies the result of iterating this map, though. What seems to be going on here?

Question 4: You have seen that at least some matrices A have special eigenvectors X such that

$$AX = aX \tag{15.6}$$

where a is some number, and aX just means scale each component of X by multiplying it by a. Assume the components of the eigenvector X are not both zero, and try solving equation (15.6) for a in terms of the the a_{ij} by the usual methods of algebra. How is this solution related to what you found numerically? How does this algebraic approach work in other examples?

15.5 Discussion

Most discussions of eigenvectors and eigenvalues begin where we ended in the last section, with equation (15.6). Frequently, though, the important thing to understand is the underlying picture, essentially Figure 15.1. In engineering applications, for example, it may happen that a process is repeated over and over, and the engineer may be concerned that nothing should grow out of control, as the vectors in Figure 15.1 seem to be doing. The design problem may be to ensure that the largest eigenvalue (in some map describing the process) be less than one in magnitude.

An amusing example of a different kind is provided by the *Fibonacci numbers* (see also Chapter 3). These numbers

$$1, 1, 2, 3, 5, 8, 13, 21, 34, \ldots$$

are the components of vectors found by iterating the map with matrix

$$A = \begin{pmatrix} 0 & 1 \\ 1 & 1 \end{pmatrix}$$

starting with $X_0 = [1, 1]$. If you find the largest eigenvalue of this A, you will have found the approximate ratio of successive Fibonacci numbers. This number is also called the "golden ratio."

We have left it for you to discover, by means of equation (15.6), that eigenvalues are solutions of a quadratic equation and so may be real or *complex*. In the complex case "largest in magnitude" must refer to the modulus of a complex number:

$$|a + bi| = \sqrt{a^2 + b^2}.$$

The situation is further complicated in this case because if there is one complex eigenvalue, then there are two, and they are of *equal* magnitude (they are complex conjugates: $a + bi$ and $a - bi$). Thus neither eigenvalue dominates in iteration, and the picture does not resemble Figure 15.1. You may also have looked at examples in which there are two *real* eigenvalues of equal magnitude. This too is a special case, with a picture unlike Figure 15.1.

Any linear algebra text will treat the eigenvalue problem. If you have a linear algebra text handy, look also at the Gauss-Seidel method for solving a linear system of equations. It is an iterative method: convergence of the method depends on whether or not the "spectral radius" of a certain matrix is less than one—spectral radius being just another name for the magnitude of the largest eigenvalue! Not all texts treat the Gauss-Seidel method. A good one in this context is

- G. Strang, *Introduction to Linear Algebra*, Wellesley-Cambridge Press, 1993 (see pp. 390-396).

15.6 Computer programs

15.6.1 *MATLAB/Octave* programs

Planar1

```
function its=planar1(A,x0,n)
%function its=planar1(A,x0,n)
%
%Estimates solution to Ax=x using iteration, for a 2x2 matrix A and
%a 2x1 vector x
%
%Inputs
%  A - a 2x2 matrix
%  x0 - a 2x1 vector
%  n - number of iterates
%Output
%  its - 2xn array of iterates, each column is an iterate`

its=zeros(2,n+1); %matrix of iterates
its(:,1)=x0; %set first column of its to initial condition
for i=1:n
its(:,i+1)=A*its(:,i); %iterate next column of its A * previous column
end; %for
plot(its(1,:),its(2,:),'kx')
```

MATLAB/Octave are matrix-oriented languages, so when we use the operation * the computer performs matrix multiplication.

Planar2

```
function []=planar2(A)
%function []=planar2(A)
%
%Plots iterates of
%
%Inputs
%  A - a 2x2 matrix
%Output
%  none, produces a plot

m=17; %number of random inital vectors
n=25; %number of iterates for each random initial vector
clf; %clear figure
hold on %hold plot so all plots show on same axes
for i=1:m
x0=[2*rand-1;2*rand-1];
planar1(A,x0,n);
end; %for
hold off %turn off plot hold
```

Chapter 16

Euclidean Algorithm for Complex Integers

16.1 Introduction

From high school you may recall that expanding our number system from the real numbers to the complex numbers enables us to solve quadratic equations like $x^2 + 1 = 0$ that have no real roots. This seems like a small, somewhat abstract problem to solve, but like the rational numbers or irrational numbers, which also can be viewed as solutions to equations extending the integers, once we have $\sqrt{-1}$ at hand, countless new worlds are open to us. Physicists use complex numbers to model and study many physical phenomena like electrical and magnetic fields, to name two simple ones. More surprising, perhaps, complex numbers are used to answer questions in number theory and geometry, areas of mathematics which, at first, seem quite far removed from them. In this chapter you will

- Calculate with complex numbers;
- Represent complex numbers geometrically;
- Learn about the *complex integers*; and
- Investigate the Euclidean algorithm for complex integers.

16.2 Complex Integers

In this chapter, we will include in our calculations a new number

$$i = \sqrt{-1}.$$

In other words, i is the number whose square equals -1: $i^2 = -1$. If we include this number, all sorts of amazing things become possible. For example, suppose we want to find *Pythagorean triples*—integer solutions to the famous equation

$$x^2 + y^2 = z^2.$$

Using $i = \sqrt{-1}$, we can factor the left-hand side of this equation into

$$(x + yi)(x - yi) = z^2.$$

One way in which this product can be a square is for both $x + iy$ and $x - iy$ to be squares. (Think about the ordinary integers.) So, we can try to find some numbers of the form $u + iv$ such that

$$x + yi = (u + vi)^2.$$

Multiplying out the right-hand side, we see that

$$x + yi = (u^2 - v^2) + (2uv)i.$$

Therefore, we must have that $x = u^2 - v^2$ and $y = 2uv$, and this means that

$$x^2 + y^2 = (u^2 - v^2)^2 + (2uv)^2 = u^4 + 2uv + v^4 = (u^2 + v^2)^2,$$

or that $z = u^2 + v^2$. So we see that using this approach and letting $z = u^2 + v^2$, we have found expressions for the Pythagorean triple $(x, y, z) = (u^2 - v^2, 2uv, u^2 + v^2)$. If we want (x, y, z) to have no common factors so that the $\gcd(x, y, z) = 1$, we will have to insist that the complex factors $(x + yi)$ and $(x - yi)$ have no common factors in the set of complex integers.[1]

[1] This example is taken from Harold Stark's introduction to quadratic fields in his book *An Introduction to Number Theory*.

By **complex integers**, we mean all numbers of the form $a + b\sqrt{-1}$ or $a+bi$ where a and b are regular integers. The complex integers $a+bi$ are also called the **Gaussian integers** because Gauss was the first person to find the right way to generalize the Euclidean algorithm for these integers. We are going to investigate the Euclidean algorithm for the complex integers. We will need to understand what a greatest common divisor is in this setting, and we will answer questions like what are the primes? and how do we recognize them?

Let's develop some notation to make our discussion easier. We denote the regular integers, also called *rational integers*, by $\mathbb{Z}$ and the new complex integers by $\mathbb{Z}[i]$; we call our familiar fractions, or ratios of integers, by the name *rational numbers* and denote them by $\mathbb{Q}$. Just as we can add, subtract, and multiply the integers in $\mathbb{Z}$ and get answers in $\mathbb{Z}$, so we can add, subtract, and multiply the complex integers in $\mathbb{Z}[i]$ and get answers in $\mathbb{Z}[i]$. These arithmetic operations are performed exactly as they would be for any binomials, but when multiplying, one simply has to remember the additional rule that $i^2 = -1$. Now, recall how we obtain all fractions in $\mathbb{Q}$ as solutions to equations of the form $ax = b$, where a and b are in $\mathbb{Z}$ and $a \neq 0$, so then $x = b/a$. In an analogous fashion we obtain the whole set of fractions of $\mathbb{Z}[i]$ by solving all equations of the form

$$(a + bi)x = c + di,$$

where $a+bi$ and $c+di$ are both complex integers and $a+bi \neq 0$. Multiplying both sides of this equation by $(a - bi)$, we see that

$$(a^2 + b^2)x = (a - bi)(c + di) = (ac + bd) + (ad - bc)i.$$

Dividing both sides by the rational integer $a^2 + b^2$, we get

$$x = \frac{ac + bd}{a^2 + b^2} + \frac{ad - bc}{a^2 + b^2}i.$$

Now notice that $\frac{ac+bd}{a^2+b^2}$ and $\frac{ad-bc}{a^2+b^2}$ are ordinary fractions in $\mathbb{Q}$; thus, x is of the form $\alpha + \beta i$, with α and β in $\mathbb{Q}$. We denote the set of all such x by $\mathbb{Q}(i)$, that is,

$$\mathbb{Q}(i) = \{\alpha + \beta i\colon \ \alpha, \beta \in \mathbb{Q}\}.$$

Like the rational numbers in $\mathbb{Q}$, all the numbers in $\mathbb{Q}(i)$ can be added, subtracted, multiplied, and divided, and, of course, both addition and multiplication are commutative. Such collections of numbers, where all the operations of ordinary arithmetic are valid, are called **fields**.

Before we describe the Euclidean algorithm for the complex integers, we need to know a little more about them. First, we picture these numbers in two dimensions on the Cartesian plane with the number $a + bi$ represented by the point (a, b). In other words, the x-axis represents the possible real components of the complex number, and the y-axis represents the complex component (or imaginary component, as it is also called). In this way, we can think of the complex integers as points in the plane with integer coordinates. The complex number $a + bi$ is, therefore, at the distance $\sqrt{a^2 + b^2}$ from the origin. We call the distance from the origin of a complex number z its **modulus** or **absolute value**, and write $|z|$. We define a new quantity, called the **norm** of a complex number z, denoted $N(z)$ to be the square of its modulus. In symbols,

$$N(a + bi) = a^2 + b^2 = (\sqrt{a^2 + b^2})^2 = |a + ib|^2.$$

We add and multiply complex integers in a very natural way. For example,

$$(1 + 2i) + (3 - 4i) = (1 + 3) + (2 - 4)i = 4 - 2i$$

and

$$\begin{aligned}(1 + 2i)(3 - 4i) &= 1(3 - 4i) + 2i(3 - 4i)\\ &= 3 - 4i + 6i - 8i^2\\ &= (3 + 8) + (6 - 4)i\\ &= 11 + 2i.\end{aligned}$$

Exercise 1: Write down the general formula for the multiplication of two complex integers. In other words, $(a + bi)(c + di) =$ what?

Exercise 2: (a) Find the norm of $(a + bi)(c + di)$ and show that $N(a + bi)N(c + di) = N((a + bi)(c + di))$. (b) If s and t are two complex numbers, $t \neq 0$, show that $N(s/t) = N(s)/N(t)$.

Exercise 3: What is the additive inverse of a complex number $a+bi$? Find the solution to the equation $2-3i+x=0$.

Exercise 4: What is the multiplicative inverse of the complex number $a+bi$? Find the solution to the equation $(2-3i)x=1$. [Hint: Look again at the solution of $(a+bi)x=c+di$ above.]

Exercise 5: When will the multiplicative inverse of an integer $a+bi$ also be a complex integer? Find all complex integers whose multiplicative inverses are also integers. Which ordinary integers have multiplicative inverses which are also integers?

Exercise 6: Find all complex integers which have norm equal to 1. Find *all* $a+bi$ such that $N(a+bi)=1$. Which ordinary integers have norm (or absolute value) equal to 1? How do these lists compare with the lists of invertible elements from Exercise 5?

Now, we are ready to consider the theorem called the complex division algorithm. Notice that, except for the use of the norm N instead of absolute value, it looks identical to the division algorithm for the ordinary integers.

Complex Division Algorithm. Given two complex integers a and b with $b \neq 0$, there are complex integers q and r such that

$$a = qb + r$$

and $0 \leq N(r) < N(b)$.

For example, if we want to divide $4+3i$ into $7+8i$, we need to find $q=q_1+q_2i$ which will be the complex integer closest to the solution to the equation

$$7+8i = (4+3i)x. \tag{16.1}$$

Let's first look at an analogue of the one-dimensional geometric argument that we made in the ordinary integer case (Chapter 3).

Recall that, there, for a given divisor b, we divided the real line into segments with endpoints $\ldots, -3b, -2b, -b, 0, b, 2b, 3b, \ldots$ and saw that the dividend a had to lie in one of those segments. The remainder r was then the distance from a to the multiple of b we chose and, hence, was always bounded in size by $|b|$.

For the complex integer case, we just generalize that same argument to make it work in a two-dimensional situation. Since we want to replace x in equation (9.1) by a nearby complex integer, we look at complex integer multiples of our divisor $4 + 3i$, as follows:

$$(4+3i)(q_1+q_2 i) = (4+3i)q_1 + (4+3i)q_2 i = (4+3i)q_1 + (-3+4i)q_2,$$

where q_1 and q_2 are ordinary integers. Note that the line segment from the origin to $4 + 3i$ has slope $3/4$ and length $|4 + 3i| = \sqrt{4^2+3^2} = 5$, while the line segment from the origin to $-3+4i$ has slope $-4/3$ and length $|-3+4i| = \sqrt{(-3)^2+4^2} = 5$. Thus, the two segments form two sides of a square.

Just as multiples of our divisor in the real case partition the real line into an infinite number of segments of length b, so do these integer multiples of $4 + 3i$ and $-3 + 4i$ divide the entire complex plane into an infinite number of squares of side $|4 + 3i|$, forming what is called a *lattice.* The vertices of the squares are called *lattice points.* Each vertex of one of these squares corresponds to a particular pair of multiples (q_1, q_2), and hence to a particular complex multiple $q_1 + q_2 i$ of our divisor $4 + 3i$. Now our dividend $7 + 8i$ must lie in one of these squares, as you can see in the Figure 16.1, and closest to at least one of the vertices. This is exactly the vertex we pick to give us the coefficients q_1 and q_2 that we are looking for in our division algorithm. Looking at the Figure 16.1, the vertex with coordinates $(8, 6)$ seems closest to our dividend (which has coordinates $(7, 8)$). The vertex $(8, 6)$ corresponds to

$$8 + 6i = 2(4 + 3i) + 0(-3 + 4i)$$

so we choose $q_1 = 2$ and $q_2 = 0$ in our example.

As specified by the theorem, the remainder defined by

$$\begin{aligned} r = r_1 + r_2 i &= (7+8i) - (4+3i)(q_1+q_2 i) \\ &= (7+81) - (4+3i)(2+0i) \\ &= -1+2i \end{aligned}$$

always has two important properties. First, because all of the coefficients on the right side of the first equation are ordinary integers r_1 and r_2, so r is a

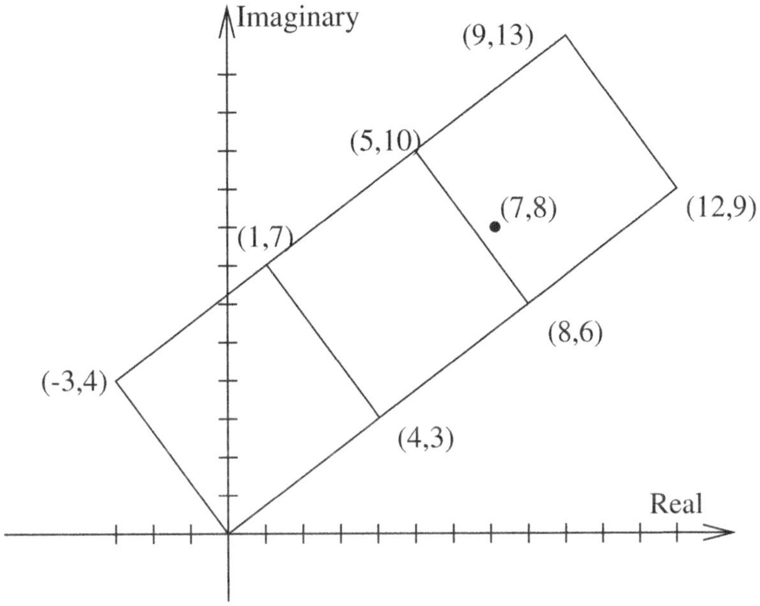

Fig. 16.1: Lattice formed by b and ib

complex integer. Second, the length of $r_1 + r_2 i$ is clearly less than the length of the side of our square; that is, in our case,

$$0 \leq |-1 + 2i| < |3 + 4i| = 5,$$

or, equivalently,

$$0 \leq N(-1 + 2i) < N(3 + 4i) = 25.$$

In general,

$$0 \leq N(r) < N(b).$$

Now let's see what happens when we look at the problem algebraically. To find the quotient q, we solve for x by dividing $7 + 8i$ by $4 + 3i$ and rationalizing the denominator:

$$x = \frac{7 + 8i}{4 + 3i} = \frac{(7 + 8i)(4 - 3i)}{(4 + 3i)(4 - 3i)} = \frac{52 + 11i}{25}.$$

Next, we let q_1 equal the integer closest to $52/25 = 2 + 2/25$, and we let q_2 equal the integer closest to $11/25$ to get $q_1 + q_2 i = 2 + 0i = 2$. Once we have the quotient, we can solve for the remainder as

$$r = a - bq = (7 + 8i) - (4 + 3i)(2)$$

and find that $r = -1 + 2i$. Notice that we are guaranteed that

$$N(r) = N(-1 + 2i) < N(4 + 3i) = N(b)$$

because

$$\frac{r}{b} = \frac{(-1 + 2i)}{(4 + 3i)} = \frac{(7 + 8i) - (4 + 3i)(2)}{(4 + 3i)} = c_1 + c_2 i = c$$

where $-0.5 \leq c_1 \leq 0.5$ and $-0.5 \leq c_2 \leq 0.5$. Since $N(c) = N(c_1 + c_2 i) \leq \frac{1}{4} + \frac{1}{4} < \frac{1}{2}$, we have that

$$0 \leq N(c) = \frac{N(r)}{N(b)} < 1$$

which implies that $0 \leq N(r) < N(b)$.

In exactly the same way that we used the division algorithm again and again for the regular integers to find the greatest common divisor, we can use this division again and again until our remainder $r = 0$, and then we will know that the last non-zero remainder is the greatest common divisor of the two complex integers we started with.

We will illustrate the procedure by continuing our example with $a = 7+8i$ and $b = 4 + 3i$.

$$\begin{aligned} 7 + 8i &= (4 + 3i)2 + (-1 + 2i) \\ 4 + 3i &= (-1 + 2i)(-2i) + i \\ -1 + 2i &= (i)(2 + i) + 0 \end{aligned}$$

Notice that at each step, the norm of the non-zero remainder is less than the norm of the divisor:

$$\begin{aligned} N(-1 + 2i) &< N(4 + 3i) \\ N(i) &< N(-1 + 2i) \end{aligned}$$

Based on this calculation, we could say that $\gcd(7+8i, 4+3i) = i$.

On the other hand, we could have chosen our q differently. Look again at Figure 16.1 to see that the lattice point corresponding to $q = 2+i$ is also a plausible choice. With that choice of q, the calculation is as follows:

$$\begin{aligned} 7+8i &= (4+3i)(2+i) + (2-2i) \\ 4+3i &= (2-2i)(2i) - i \\ 2-2i &= (-i)(-2-2i) + 0 \end{aligned}$$

Again, we have "proper" remainders at every step:

$$\begin{aligned} N(2-2i) &< N(4+3i) \\ N(-i) &< N(2-2i) \end{aligned}$$

In this case, we would like to say that $\gcd(7+8i, 4+3i) = -i$.

On still another hand, we could have worked with the same q in the following manner:

$$\begin{aligned} 7+8i &= (4+3i)(2+i) + (2-2i) \\ 4+3i &= (2-2i)(i) + (2+i) \\ 2-2i &= (2+i)(-i) + 1 \\ 2+i &= 1(2+i) + 0 \end{aligned}$$

The remainders are proper in this case too:

$$\begin{aligned} N(2-2i) &< N(4+3i) \\ N(2+i) &< N(2-2i) \\ N(1) &< N(2+i) \end{aligned}$$

From this calculation, we would like to say that $\gcd(7+8i, 4+3i) = 1$.

Exercise 7: In light of your answer to Exercise 6 above, what is the greatest common divisor of $a = 7+8i$ and $b = 4+3i$, and what is going on here?

Exercise 8: In general, how many different choices for each quotient are there which still allow the norm of the remainder to be smaller than the norm of the divisor? Consider the possibilities for $q_1 + q_2 i$. When will there be only one choice for $q_1 + q_2 i$? When will there be 2, 3, 4 or more?

16.3 Questions and discussion

Here is the pseudocode for a program which implements the complex Euclidean algorithm. It is an adaptation of the program EUCLID1 from Chapter 3. The actual program works with the real and imaginary components of each complex number. Here we write the complex numbers only.

Program outline: EuclidC

```
Input: complex integers a and b
Output: the steps of the algorithm, greatest common divisor
        and the number steps to reach it

as := a
bs := b
% as and bs are initially 1x1 dynamic arrays which will
% increase in length as the program progresses.

i := 1
WHILE bs(i) ~= 0
   as(i+1) := bs(i)
   x := as(i)/bs(i)
   q := x with its real and imaginary parts rounded
   bs(i+1) := as(i)- q * bs(i)
   i := i + 1
END WHILE

k := length(as) %the number of steps
thegcd := the last entry  of as
OUTPUT as, bs, thegcd, k
```

Try to answer the same questions for this version of the Euclidean algorithm that you answered for the algorithm for ordinary integers.

Question 1: Can you find a bound on the number of steps that the algorithm takes to finish. The norm $N(b)$, is an obvious rough bound. Can you find a better bound?

Question 2: Two complex integers are **relatively prime** if their greatest common divisor is ± 1 or $\pm i$. First, explain, by analogy to the ordinary inegers, why this definition makes sense. What is the probability that two randomly chosen complex integers are relatively prime? You can estimate this probability by selecting a large number, m say, of randomly chosen pairs of complex integers (with coordinates between 0 and N, say). The number of relatively prime pairs divided by m will be an estimate of the true proportion of relatively prime pairs whose coordinates are between 0 and N.

The program PROPORTION below, which is adapted from EUCLID4 in Chapter 3, will enable you to investigate Question 2.

Program outline: Proportion

```
Input: N > 0 and number m of pairs of complex integers
       with coordinates between 0 and N
Output: Proportion of the m pairs that are relatively prime

s := 0  % s counts relatively prime pairs
FOR i = 1 TO m
  Randomly select a pair a,b of complex integers
  RUN EuclidC
  IF thegcd==1 OR thegcd==-1 OR thegcd==I OR thegcd==-I
    s=s+1 %the numbers where relatively prime
  END IF
END WHILE
prop := s/m  ! Proportion of relatively prime pairs
OUTPUT prop
```

Question 3: What are the prime numbers in the complex integers? Is 2 still prime? Are 3, 5, 7, and 11? Can you adapt the sieve method to list all primes with norm less than or equal to $N(11)$?

References:

[1] W.K. Buhler. *Gauss—A Biographical Study.* Springer-Verlag, 1981.

[2] Harvey Cohn. *Advanced Number Theory.* Dover, 1982.

[3] Harold M. Stark. *An Introduction to Number Theory.* MIT Press, 1978 (originally published by Markham in 1970).

16.4 Computer Programs

16.4.1 *MATLAB/Octave* programs

Program: EuclidC

```
function [as,bs,thegcd,k]=euclidc(a,b)
%function [as,bs,thegcd,k]=euclidc(a,b)
%
%Applies complex Euclidean algorithm to complex integers a and b
%
%Inputs
%  a - complex integer to apply Eucliden Algorithm to
%  b - other integer to apply Eucliden Algorithm to
%      Algorithm assumes a>b, if not code will swap a and b
%Output
%  as - vector of sequence of a values
%  bs - vector of sequence of b values
%  thegcd - the gcd as computed by algorithm
%  k - number of steps required

as=a;  %initialize "as" vector with b
bs=b;  %initialize "bs" vector with b

i=1; %itialize variable to count step
while bs(i)~=0 %repeat until b is 0
   as(i+1)=bs(i);  %next a is previous b
```

```
   x=as(i)/bs(i);
   q=round(real(x))+round(imag(x))*I; %closest complex integer to a/b
   bs(i+1)=as(i)-bs(i)*q; %next b is remainder of a/b
   i=i+1; %increment step count
end; %while
k = length(as); %number of steps required
thegcd = as(end); %the gcd is the last entry of the as vector
```

Program: Proportion

```
function prop=proportion(N,m)
%function prop=proportion(N,m)
%
%Applies Euclidean algorithm to m pairs of random integers each
%with real and imaginary parts bounded by N and determines the
%proportion of those pairs that are realtively prime.
%
%Inputs
%  N - maximum absolute value of real and imaginary parts of
%      randomly selected complex integers
%  m - number of random pairs to apply Euclidean algorithm to
%Output
%  prop - proportion of randomly selected pairs that were
%         relatively prime

s = 0; %initialize count of relatively prime pairs
for i=1:m  %repeat m times
   a=floor(1 + N * rand) +  floor(1 + N * rand) * I;
   b=floor(1 + N * rand) +  floor(1 + N * rand) * I;
   [as,bs,thegcd,k]=euclidc(a,b); %apply Euclid2 code
   if thegcd==I || thegcd==-I || thegcd==1 || thegcd==-1
      s=s+1; %the numbers were relatively prime
   end; %if
end; %for
prop=s/m; %divide s by m to get proportion
```

16.4.2 *Mathematica* programs

Single pair of complex integers (EuclidC)

"EuclidC" computes the GCD of two complex integers. The output shows and counts all intermediate steps. The code is identical to that for ordinary integers.

```
EuclidC[a_, b_]:=Module[{c = a, d = b, ctr = 1}, {c, d} = Sort[{c, d}];
    Print["Step ", ctr, ":\t ", c, "\t", d];
    While[c ≠ 0, {c, d} = {Mod[d, c], c};
        ctr++;
        Print["Step ", ctr, ": \t", c, "\t", d]]]
```

Here is an example, the GCD of 2+16I and 1+9I (note the use of "I" to input the imaginary unit, though the output of the imaginary unit is a stylized "i"):

EuclidC[2 + 16*I*, 1 + 9*I*]

Step 1:	$1 + 9i$	$2 + 16i$
Step 2:	$-2i$	$1 + 9i$
Step 3:	$1 + i$	$-2i$
Step 4:	0	$1 + i$

Here is the first example from the chapter:

EuclidC[7 + 8*I*, 4 + 3*I*]

Step 1:	$4 + 3i$	$7 + 8i$

Step 2:	$-1+2i$	$4+3i$
Step 3:	i	$-1+2i$
Step 4:	0	i

Counting relatively prime pairs (Proportion)

This program uses *Mathematica*'s built-in GCD function, and so the GCD's all have positive real component. Thus two numbers are relatively prime precisely when their GCD equals 1. The graph is less useful in this case, and so the program just outputs a table.

```
RCI[n_Integer]:=RandomInteger[{0, n}] + RandomInteger[{0, n}]i;
Prop[n_Integer, M_Integer]:=
    (Print["GCD distribution for ", M,
    " pairs of random complex integers whose components are chosen less than "
    , n];
store = {};
Do[store = Append[store, GCD[RCI[n], RCI[n]]], {M}];
Print[TableForm[Sort[Tally[store]], TableHeadings → {None, {"GCD", "Frequency"}}]])
```

Here is an example.

```
Prop[100, 1000]
```

GCD distribution for 1000 pairs of random complex integers whose components are chosen less than 100

GCD	Frequency
1	667
$1+i$	177
$1+2i$	24
$1+3i$	2
$1+4i$	2
$1+6i$	2
2	50
$2+i$	18
$2+2i$	19
$2+3i$	3
$2+4i$	1
$2+6i$	1
$2+8i$	1
3	7
$3+i$	5
$3+2i$	4
$3+3i$	2
$3+4i$	2
$3+5i$	1
$4+i$	2
$4+2i$	1
5	1
$5+i$	1
$5+5i$	2
6	1
$6+2i$	1
8	1
$14+7i$	1
$57+18i$	1

Index

Made in the USA
Monee, IL
08 November 2026

67377453R00177